THE INEVITABLE

ALSO BY KEVIN KELLY

Out of Control:
The New Biology of Machines, Social Systems,
and the Economic World

New Rules for the New Economy:
10 Radical Strategies for a Connected World

Asia Grace

What Technology Wants

Cool Tools:
A Catalog of Possibilities

THE
INEVITABLE

UNDERSTANDING THE **12 TECHNOLOGICAL**
FORCES THAT WILL SHAPE OUR FUTURE

KEVIN KELLY

VIKING

VIKING
An imprint of Penguin Random House LLC
375 Hudson Street
New York, New York 10014
penguin.com

ISBN 9780525428084 (hardcover)
ISBN 9780698183650 (ebook)

Printed in the United States of America
10   9   8   7   6

Set in Minion Pro with Aeonis LT Pro
Designed by Daniel Lagin

CONTENTS

INTRODUCTION 1

1. BECOMING 9

2. COGNIFYING 29

3. FLOWING 61

4. SCREENING 85

5. ACCESSING 109

6. SHARING 135

7. FILTERING 165

8. REMIXING 193

CONTENTS

9. INTERACTING   211

10. TRACKING   237

11. QUESTIONING   269

12. BEGINNING   291

ACKNOWLEDGMENTS   299

NOTES   301

INDEX   317

THE INEVITABLE

INTRODUCTION

When I was 13, my father took me to visit a computer trade show in Atlantic City, New Jersey. It was 1965 and he was excited by these room-size machines made by the smartest corporations in America, such as IBM. My father believed in progress, and these very first computers were glimpses of the future he imagined. But I was very *unimpressed*—a typical teenager. The computers filling the cavernous exhibit hall were boring. There was nothing to see except acres of static rectangular metal cabinets. Not a single flickering screen anywhere. No speech input, or output. The only thing these computers could do was print out rows and rows of gray numbers on folded paper. I knew a lot about computers from my avid reading of science fiction, and these were not *real* computers.

In 1981 I got my hands on an Apple II computer in a science lab at the University of Georgia, where I worked. Even though it had a tiny green and black screen that could display text, I was not impressed by this computer either. It could do typing better than a typewriter, and it was a whiz with graphing numbers and keeping track of data, but it was not a *real* computer. It was not rearranging my life.

My opinions totally changed a few months later when I plugged the same Apple II into a phone line with a modem. Suddenly everything was different. There was an emerging universe on the other side of the phone jack, and it was huge, almost infinite. There were online bulletin boards, experimental teleconferences, and this place called the internet. The portal through the phone line opened up something both vast and at the same time human scaled. It felt organic and fabulous. It connected people and machines in a personal way. I could feel my life jumping up to another level.

Looking back, I think the computer age did not really start until this moment, when computers merged with the telephone. Stand-alone computers were inadequate. All the enduring consequences of computation did not start until the early 1980s, that moment when computers married phones and melded into a robust hybrid.

In the three decades since then, this technological convergence between communication and computation has spread, sped up, blossomed, and evolved. The internet/web/mobile system has moved from the fringes of society (where it was pretty much ignored in 1981) to the center stage of our modern global society. In the past 30 years the social economy based on this technology has had its ups and downs and seen its heroes come and go, but it is very clear there have been large-scale trends governing what has happened.

These broad historical trends are crucial because the underlying conditions that birthed them are still active and developing, which strongly suggests that these trends will continue to increase in the next few decades. There is nothing on the horizon to decrease them. Even the forces we might think could derail them, like crime, war, or our own excesses, also follow these emerging patterns. In this book I describe a dozen of these inevitable technological forces that will shape the next 30 years.

"Inevitable" is a strong word. It sends up red flags for some people because they object that nothing is inevitable. They claim that human willpower and purpose can—and should!—deflect, overpower, and control any mechanical trend. In their view, "inevitability" is a free will cop-out we surrender to. When the notion of the inevitable is forged with fancy technology, as I do here, the objections to a preordained destiny are even more fierce and passionate. One definition of "inevitable" is the final outcome in the classic rewinding thought experiment. If we rewound the tape of history back to the beginning of time and reran our civilization from the start again and again, a strong version of inevitability says that, no matter how many times we reran it, every time we end up with teenagers tweeting every five minutes in 2016. That's not what I mean.

I mean inevitable in a different way. There is bias in the nature of technology that tilts it in certain directions and not others. All things being equal, the physics and mathematics that rule the dynamics of technology tend to favor certain behaviors. These tendencies exist primarily in the aggregate forces that shape the general contours of technological forms and do not govern specifics or particular instances. For example, the form of an internet—a network of networks spanning the globe—was inevitable, but the specific kind of internet we chose to have was not. The internet could have been commercial rather than nonprofit, or a national system instead of international, or it could have been secret instead of public. Telephony—long-distance electrically transmitted voice messages—was inevitable, but the iPhone was not. The generic form of a four-wheeled vehicle was inevitable, but SUVs were not. Instant messaging was inevitable, but tweeting every five minutes was not.

Tweeting every five minutes is not inevitable in another way. We are morphing so fast that our ability to invent new things outpaces the rate we can civilize them. These days it takes us a decade after a technology appears to develop a social consensus on what it means and what etiquette we need to tame it. In another five years we'll find a polite place for twittering, just as we figured out what to do with cell phones ringing everywhere. (Use silent vibrators.) Just like that, this initial response will disappear quickly and we'll see it was neither essential nor inevitable.

The kind of inevitability I am speaking of here in the digital realm is the result of momentum. The momentum of an ongoing technological shift. The strong tides that shaped digital technologies for the past 30 years will continue to expand and harden in the next 30 years. These apply to not just North America, but to the entire world. Throughout this book I use examples from the United States because readers will be more familiar with them, but for each I could have easily found a corresponding example in India, Mali, Peru, or Estonia. The true leaders in digital money, for example, are in Africa and Afghanistan, where e-money is sometimes the only functioning currency. China is way ahead of everyone else in developing sharing applications on mobile. But while culture can advance or retard the expression, the underlying forces are universal.

After living online for the past three decades, first as a pioneer in a rather wild empty quarter, and then later as a builder who constructed parts of this new continent, my confidence in this inevitability is based on the depth of these technological changes. The daily glitter of high-tech novelty rides upon slow currents. The roots of the digital world are anchored in the physical needs and natural tendencies of bits, information, and networks. No matter what geography, no matter what companies, no matter what politics, these fundamental ingredients of bits and networks will hatch similar results again and again. Their inevitability stems from their basic physics. In this book I endeavor to expose these roots of digital technology because from them will issue the enduring trends in the next three decades.

Not all of this shift will be welcomed. Established industries will topple because their old business models no longer work. Entire occupations will disappear, together with some people's livelihoods. New occupations will be born and they will prosper unequally, causing envy and inequality. The continuation and extension of the trends I outline will challenge current legal assumptions and tread on the edge of outlaw—a hurdle for law-abiding citizens. By its nature, digital network technology rattles international borders because it is borderless. There will be heartbreak, conflict, and confusion in addition to incredible benefits.

Our first impulse when we confront extreme technology surging forward in this digital sphere may be to push back. To stop it, prohibit it, deny it, or at least make it hard to use. (As one example, when the internet made it easy to copy music and movies, Hollywood and the music industry did everything they could to stop the copying. To no avail. They succeeded only in making enemies of their customers.) Banning the inevitable usually backfires. Prohibition is at best temporary, and in the long counterproductive.

A vigilant, eyes-wide-open embrace works much better. My intent in this book is to uncover the roots of digital change so that we can embrace them. Once seen, we can work *with* their nature, rather than struggle against it. Massive copying is here to stay. Massive tracking and total surveillance is here to stay. Ownership is shifting away. Virtual reality is becoming real. We can't stop artificial intelligences and robots from improving, creating new businesses, and taking our current jobs. It may be against our initial impulse, but we should embrace the perpetual remixing of these technologies. Only by working with these technologies, rather than trying to thwart them, can we gain the best of what they have to offer. I don't mean to keep our hands off. We need to manage these emerging inventions to prevent actual (versus hypothetical) harms, both by legal and technological means. We need to civilize and tame new inventions in their particulars. But we can do that only with deep engagement, firsthand experience, and a vigilant acceptance. We can and should regulate Uber-like taxi services, as an example, but we can't and shouldn't attempt to prohibit the inevitable decentralization of services. These technologies are not going away.

Change is inevitable. We now appreciate that *everything* is mutable and undergoing change, even though much of this alteration is imperceptible. The highest mountains are slowly wearing away under our feet, while every animal and plant species on the planet is morphing into something different in ultra slow motion. Even the eternal shining sun is fading on an astronomical schedule, though we will be long gone when it does. Human culture, and biology too, are part of this imperceptible slide toward something new.

At the center of every significant change in our lives today is a technology of some sort. Technology is humanity's accelerant. Because of technology everything we make is always in the process of becoming. Every kind of thing is becoming something else, while it churns from "might" to "is." All is flux. Nothing is finished. Nothing is done. This never-ending change is the pivotal axis of the modern world.

Constant flux means more than simply "things will be different." It means processes—the engines of flux—are now more important than products. Our greatest invention in the past 200 years was not a particular gadget or tool but the invention of the scientific process itself. Once we invented the scientific method, we could immediately create thousands of other amazing things we could have never discovered any other way. This methodical process of constant change and improvement was a million times better than inventing any particular product, because the process generated a million new products over the centuries since we invented it. Get the ongoing process right and it will keep generating ongoing benefits. In our new era, processes trump products.

This shift toward processes also means ceaseless change is the fate for everything we make. We are moving away from the world of fixed nouns and toward a world of fluid verbs. In the next 30 years we will continue to take solid things—an automobile, a shoe—and turn them into intangible verbs. Products will become services and processes. Embedded with high doses of technology, an automobile becomes a transportation service, a continuously updated sequence of materials rapidly adapting to customer usage, feedback, competition, innovation, and wear. Whether it is a driverless car or one you drive, this transportation service is packed with flexibility, customization, upgrades, connections, and new benefits. A shoe, too, is no longer a finished product, but an endless process of reimagining our extended feet, perhaps with disposable covers, sandals that morph as you walk, treads that shift, or floors that act as shoes. "Shoeing" becomes a service and not a noun. In the intangible digital realm, nothing is static or fixed. Everything is becoming.

Upon this relentless change all the disruptions of modernity ride. I've waded through the myriad technological forces erupting into the present and I've sorted their change into 12 verbs, such as *accessing*, *tracking*, and *sharing*. To be more accurate, these are not just verbs, but *present participles*, the grammatical form that conveys *continuous action*. These forces are accelerating actions.

Each of these 12 continuous actions is an ongoing trend that shows all evidence of continuing for at least three more decades. I call these metatrends "inevitable" because they are rooted in the nature of technology, rather than in the nature of society. The character of the verbs follows the biases present in the new technologies, a bias all technologies share. While we creators have much choice and responsibility in steering technologies, there is also much about a technology that is outside of our control. Particular technological processes will inherently favor particular outcomes. For instance, industrial processes (like steam engines, chemical plants, dams) favor temperatures and pressures outside of human comfort zones, and digital technologies (computers, internet, apps) favor cheap ubiquitous duplication. The bias toward high pressure/high temperature for industrial processes steers places of manufacturing away from humans and toward large-scale, centralized factories, regardless of culture, background, or politics. The bias toward cheap ubiquitous copies in digital technologies is independent of nationality, economic momentum, or human desire, and it steers the technology toward social ubiquity; the bias is baked into the nature of digital bits. In both of these examples, we can get the most from the technologies when we "listen" to the direction the technologies lean, and bend our expectations, regulations, and products to these fundamental tendencies within that technology. We'll find it easier to manage the complexities, optimize the benefits, and reduce the harm of particular technologies when we align our uses with their biased trajectory. The purpose of this book is to gather those tendencies now operating in the newest technologies and to lay their trajectories out before us.

These organizing verbs represent the metachanges in our culture for the foreseeable near future. They are broad strokes already operating in the world today. I make no attempt to predict which specific products will prevail next year or the next decade, let alone which companies will triumph. These specifics are decided by fads, fashion, and commerce, and are wholly unpredictable. But the general trends of the products and services in 30 years are currently visible. Their basic forms are rooted in directions generated by emerging technologies now on their way to ubiquity. This wide, fast-moving system of technology bends the culture subtly, but steadily, so it amplifies the following forces: Becoming, Cognifying, Flowing, Screening, Accessing, Sharing, Filtering, Remixing, Interacting, Tracking, Questioning, and then Beginning.

While I devote a chapter to each motion, they are not discrete verbs operating in solo. Rather they are highly overlapping forces, each co-dependent upon and mutually accelerating the others. It becomes difficult to speak of one without referring to the others at the same time. Increased *sharing* both encourages increased *flowing* and depends upon it. *Cognifying* requires *tracking*. *Screening* is inseparable from *interacting*. The verbs themselves are *remixed*, and all of these actions are variations on the process of *becoming*. They are a unified field of motion.

These forces are trajectories, not destinies. They offer no predictions of where we end up. They tell us simply that in the near future we are headed inevitably in these directions.

1

BECOMING

I t's taken me 60 years, but I had an epiphany recently: Everything, without exception, requires additional energy and order to maintain itself. I knew this in the abstract as the famous second law of thermodynamics, which states that everything is falling apart slowly. This realization is not just the lament of a person getting older. Long ago I learned that even the most inanimate things we know of—stone, iron columns, copper pipes, gravel roads, a piece of paper—won't last very long without attention and fixing and the loan of additional order. Existence, it seems, is chiefly maintenance.

What has surprised me recently is how unstable even the intangible is. Keeping a website or a software program afloat is like keeping a yacht afloat. It is a black hole for attention. I can understand why a mechanical device like a pump would break down after a while—moisture rusts metal, or the air oxidizes membranes, or lubricants evaporate, all of which require repair. But I wasn't thinking that the nonmaterial world of bits would also degrade. What's to break? Apparently everything.

Brand-new computers will ossify. Apps weaken with use. Code

corrodes. Fresh software just released will immediately begin to fray. On their own—nothing you did. The more complex the gear, the more (not less) attention it will require. The natural inclination toward change is inescapable, even for the most abstract entities we know of: bits.

And then there is the assault of the changing digital landscape. When everything around you is upgrading, this puts pressure on your digital system and necessitates maintenance. You may not want to upgrade, but you must because everyone else is. It's an upgrade arms race.

I used to upgrade my gear begrudgingly (why upgrade if it still works?) and at the last possible moment. You know how it goes: Upgrade this and suddenly you need to upgrade that, which triggers upgrades everywhere. I would put it off for years because I had the experiences of one "tiny" upgrade of a minor part disrupting my entire working life. But as our personal technology is becoming more complex, more codependent upon peripherals, more like a living ecosystem, *delaying* upgrading is even more disruptive. If you neglect ongoing minor upgrades, the change backs up so much that the eventual big upgrade reaches traumatic proportions. So I now see upgrading as a type of hygiene: You do it regularly to keep your tech healthy. Continual upgrades are so critical for technological systems that they are now automatic for the major personal computer operating systems and some software apps. Behind the scenes, the machines will upgrade themselves, slowly changing their features over time. This happens gradually, so we don't notice they are "becoming."

We take this evolution as normal.

Technological life in the future will be a series of endless upgrades. And the rate of graduations is accelerating. Features shift, defaults disappear, menus morph. I'll open up a software package I don't use every day expecting certain choices, and whole menus will have disappeared.

No matter how long you have been using a tool, endless upgrades make you into a newbie—the new user often seen as clueless. In this era of "becoming," everyone becomes a newbie. Worse, we will be newbies forever. That should keep us humble.

That bears repeating. All of us—every one of us—will be endless newbies in the future simply trying to keep up. Here's why: First, most of the important technologies that will dominate life 30 years from now have not yet been invented, so naturally you'll be a newbie to them. Second, because the new technology requires endless upgrades, you will remain in the newbie state. Third, because the cycle of obsolescence is accelerating (the average lifespan of a phone app is a mere 30 days!), you won't have time to master anything before it is displaced, so you will remain in the newbie mode forever. Endless Newbie is the new default for everyone, no matter your age or experience.

———

If we are honest, we must admit that one aspect of the ceaseless upgrades and eternal becoming of the technium is to make holes in our heart. One day not too long ago we (all of us) decided that we could not live another day unless we had a smartphone; a dozen years earlier this need would have dumbfounded us. Now we get angry if the network is slow, but before, when we were innocent, we had no thoughts of the network at all. We keep inventing new things that make new longings, new holes that must be filled.

Some people are furious that our hearts are pierced this way by the things we make. They see this ever-neediness as a debasement, a lowering of human nobility, the source of our continual discontentment. I agree that technology is the source. The momentum of technologies pushes us to chase the newest, which are always disappearing beneath the advent of the next newer thing, so satisfaction continues to recede from our grasp.

But I celebrate the never-ending discontentment that technology brings. We are different from our animal ancestors in that we are not content to merely survive, but have been incredibly busy making up new itches that we have to scratch, creating new desires we've never had before. This discontent is the trigger for our ingenuity and growth.

We cannot expand our self, and our collective self, without making holes in our heart. We are stretching our boundaries and widening the small container that holds our identity. It can be painful. Of course, there will be rips and tears. Late-night infomercials and endless web pages of about-to-be-obsolete gizmos are hardly uplifting techniques, but the path to our enlargement is very prosaic, humdrum, and everyday. When we imagine a better future, we should factor in this constant discomfort.

––––––––

A world without discomfort is utopia. But it is also stagnant. A world perfectly fair in some dimensions would be horribly unfair in others. A utopia has no problems to solve, but therefore no opportunities either.

None of us have to worry about these utopia paradoxes, because utopias never work. Every utopian scenario contains self-corrupting flaws. My aversion to utopias goes even deeper. I have not met a speculative utopia I would want to live in. I'd be bored in utopia. Dystopias, their dark opposites, are a lot more entertaining. They are also much easier to envision. Who can't imagine an apocalyptic last-person-on-earth finale, or a world run by robot overlords, or a megacity planet slowly disintegrating into slums, or, easiest of all, a simple nuclear Armageddon? There are endless possibilities of how the modern civilization collapses. But just because dystopias are cinematic and dramatic, and much easier to imagine, that does not make them likely.

The flaw in most dystopian narratives is that they are not sustainable. Shutting down civilization is actually hard. The fiercer the disaster, the faster the chaos burns out. The outlaws and underworlds that seem so exciting at "first demise" are soon taken over by organized crime and militants, so that lawlessness quickly becomes racketeering and, even quicker, racketeering becomes a type of corrupted government—all to maximize the income of the bandits. In a sense, greed cures anarchy. Real dystopias are more like the old Soviet Union rather than *Mad Max*: They are stiflingly bureaucratic rather than lawless. Ruled by fear, their society is hobbled except for the benefit of a few, but, like the sea pirates two centuries ago, there is far more law and order than appears. In fact, in real broken societies, the outrageous outlawry we associate with dystopias is not permitted. The big bandits keep the small bandits and dystopian chaos to a minimum.

However, neither dystopia nor utopia is our destination. Rather, technology is taking us to *protopia*. More accurately, we have already arrived in protopia.

Protopia is a state of becoming, rather than a destination. It is a process. In the protopian mode, things are better today than they were yesterday, although only a little better. It is incremental improvement or mild progress. The "pro" in protopian stems from the notions of process and progress. This subtle progress is not dramatic, not exciting. It is easy to miss because a protopia generates almost as many new problems as new benefits. The problems of today were caused by yesterday's technological successes, and the technological solutions to today's problems will cause the problems of tomorrow. This circular expansion of both problems and solutions hides a steady accumulation of small net benefits over time. Ever since the Enlightenment and the invention of science, we've managed to create a tiny bit more than we've destroyed each year. But that few percent positive difference is compounded over decades into what we might call civilization. Its benefits never star in movies.

Protopia is hard to see because it is a becoming. It is a process that is constantly changing how other things change, and, changing itself, is mutating and growing. It's difficult to cheer for a soft process that is shape-shifting. But it is important to see it.

Today we've become so aware of the downsides of innovations, and so disappointed with the promises of past utopias, that we find it hard to believe even in a mild protopian future—one in which tomorrow will be a little better than today. We find it very difficult to imagine any kind of future at all that we desire. Can you name a single science fiction future on this planet that is both plausible and desirable? (*Star Trek* doesn't count; it's in space.)

There is no happy flying-car future beckoning us any longer. Unlike the last century, nobody wants to move to the distant future. Many dread it. That makes it hard to take the future seriously. So we're stuck in the short now, a present without a generational perspective. Some have adopted the perspective of believers in a Singularity who claim that imagining the future in 100 years is technically impossible. That makes us future-blind. This future-blindness may simply be the inescapable affliction of our modern world. Perhaps at this stage in civilization and technological advance, we enter into a permanent and ceaseless present, without past or future. Utopia, dystopia, and protopia all disappear. There is only the Blind Now.

The other alternative is to embrace the future and its becoming. The future we are aimed at is the product of a process—a becoming—that we can see right now. We can embrace the current emerging shifts that will become the future.

The problem with constant becoming (especially in a protopian crawl) is that unceasing change can blind us to its incremental changes. In constant motion we no longer notice the motion. Becoming is thus a self-cloaking action often seen only in retrospect. More important, we tend to see new things from the frame of the old. We extend our current perspective to the future, which in fact distorts the new to fit into what we already know. That is why the first movies were filmed like theatrical plays and the first VRs shot like movies. This shoehorning is not always bad. Storytellers exploit this human reflex in order to relate the new to the old, but when we are trying to discern what will happen in front of us, this habit can fool us. We have great difficulty perceiving change that is happening right now. Sometimes its apparent trajectory seems impossible, implausible, or ridiculous, so we dismiss it. We are constantly surprised by things that have been happening for 20 years or longer.

I am not immune from this distraction. I was deeply involved in the birth of the online world 30 years ago, and a decade later the arrival of the web. Yet at every stage, what was becoming was hard to see in the moment. Often it was hard to believe. Sometimes we didn't see what was becoming because we didn't want it to happen that way.

We don't need to be blind to this continuous process. The rate of change in recent times has been unprecedented, which caught us off guard. But now we know: We are, and will remain, perpetual newbies. We need to believe in improbable things more often. Everything is in flux, and the new forms will be an uncomfortable remix of the old. With effort and imagination we can learn to discern what's ahead more clearly, without blinders.

Let me give you an example of what we can learn about our future from the very recent history of the web. Before the graphic Netscape browser illuminated the web in 1994, the text-only internet did not exist for most people. It was hard to use. You needed to type code. There were no pictures. Who wanted to waste time on something so boring? If it was acknowledged at all in the 1980s, the internet was dismissed as either corporate email (as exciting as a necktie) or a clubhouse for teenage boys. Although it did exist, the internet was totally ignored.

Any promising new invention will have its naysayers, and the bigger the promises, the louder the nays. It's not hard to find smart people saying stupid things about the web/internet on the morning of its birth. In late 1994, *Time* magazine explained why the internet would never go mainstream: "It was not designed for doing commerce, and it does not gracefully accommodate new arrivals." Wow! *Newsweek* put the doubts more bluntly in a February 1995 headline: "The Internet? Bah!" The article was written by an astrophysicist and network expert, Cliff Stoll, who argued that online shopping and online communities were an unrealistic fantasy that betrayed common sense. "The truth is no online database will replace your newspaper," he claimed. "Yet Nicholas Negroponte, director of the MIT Media Lab, predicts that we'll soon buy books and newspapers straight over the Internet. Uh, sure." Stoll captured the prevailing skepticism of a digital world full of "interacting libraries, virtual communities, and electronic commerce" with one word: "baloney."

This dismissive attitude pervaded a meeting I had with the top leaders of ABC in 1989. I was there to make a presentation to the corner-office crowd about this "Internet Stuff." To their credit, the executives of ABC realized something was happening. ABC was one of the top three mightiest television networks in the world; the internet at that time was a mere mosquito in comparison. But people living on the internet (like me) were saying it could disrupt their business. Still, nothing I could tell them would convince them that the internet was not marginal, not just typing, and, most emphatically, not just teenage boys. But all the sharing, all the free stuff seemed too impossible to business executives. Stephen Weiswasser, a senior VP at ABC, delivered the ultimate put-down: "The Internet will be the CB radio of the '90s," he told me, a charge he later repeated to the press. Weiswasser summed up ABC's argument for ignoring the new medium: "You aren't going to turn passive consumers into active trollers on the internet."

I was shown the door. But I offered one tip before I left. "Look," I said. "I happen to know that the address abc.com has not been registered. Go down to your basement, find your most technical computer geek, and have him register abc.com immediately. Don't even think about it. It will be a good thing to do." They thanked me vacantly. I checked a week later. The domain was still unregistered.

While it is easy to smile at the sleepwalkers in TV land, they were not the only ones who had trouble imagining an alternative to couch potatoes. *Wired* magazine did too. I was a co-founding editor of *Wired*, and when I recently reexamined issues of *Wired* from the early 1990s (issues that I'd proudly edited), I was surprised to see them touting a future of high production-value content—5,000 always-on channels and virtual reality, with a sprinkling of bits of the Library of Congress. In fact, *Wired* offered a vision nearly identical to that of internet wannabes in the broadcast, publishing, software, and movie industries, like ABC. In this official future, the web was basically TV that worked. With a few clicks you could choose any of 5,000 channels of relevant material to browse, study, or watch, instead of the TV era's five channels. You could jack into any channel you wanted from "all sports all the time" to the saltwater aquarium channel. The only uncertainty was, who would program it all? *Wired* looked forward to a constellation of new media upstarts like Nintendo and Yahoo! creating the content, not old-media dinosaurs like ABC.

Problem was, content was expensive to produce, and 5,000 channels of it would be 5,000 times as costly. No company was rich enough, no industry large enough to carry off such an enterprise. The great telecom companies, which were supposed to wire up the digital revolution, were paralyzed by the uncertainties of funding the net. In June 1994, David Quinn of British Telecom admitted to a conference of software publishers, "I'm not sure how you'd make money out of the internet." The immense sums of money supposedly required to fill the net with content sent many technocritics into a tizzy. They were deeply concerned that cyberspace would become cyburbia—privately owned and operated.

The fear of commercialization was strongest among hard-core programmers who were actually building the web: the coders, Unix weenies, and selfless volunteer IT folk who kept the ad hoc network running. The techy administrators thought of their work as noble, a gift to humanity. They saw the internet as an open commons, not to be undone by greed or commercialization. It's hard to believe now, but until 1991 commercial enterprise on the internet was strictly prohibited as an unacceptable use. There was no selling, no ads. In the eyes of the National Science Foundation (which ran the internet backbone), the internet was funded for research, not commerce. In what seems remarkable naiveté now, the rules favored public institutions and forbade "extensive use for private or personal business." In the mid-1980s I was involved in shaping the WELL, an early text-only online system. We struggled to connect our private WELL network to the emerging internet because we were thwarted, in part, by the NSF's "acceptable use" policy. The WELL couldn't prove its users would *not* conduct commercial business on the internet, so we were not allowed to connect. We were all really blind to what was becoming.

This anticommercial attitude prevailed even in the offices of *Wired*. In 1994, during the first design meetings for *Wired*'s embryonic website, *HotWired*, our programmers were upset that the innovation we were cooking up—the first ever click-through ad banner—subverted the great social potential of this new territory. They felt the web was hardly out of diapers, and already they were being asked to blight it with billboards and commercials. But prohibiting the flow of money within this emerging parallel civilization was crazy. Money in cyberspace was inevitable.

That was a small misperception compared with the bigger story we all missed.

Computing pioneer Vannevar Bush outlined the web's core idea—hyperlinked pages—way back in 1945, but the first person to try to build out the concept was a freethinker named Ted Nelson, who envisioned his own scheme in 1965. However, Nelson had little success connecting digital bits on a useful scale, and his efforts were known only to an isolated group of disciples.

At the suggestion of a computer-savvy friend, I got in touch with Nelson in 1984, a decade before the first websites. We met in a dark dockside bar in Sausalito, California. He was renting a houseboat nearby and had the air of someone with time on his hands. Folded notes erupted from his pockets and long strips of paper slipped from overstuffed notebooks. Wearing a ballpoint pen on a string around his neck, he told me— way too earnestly for a bar at four o'clock in the afternoon—about his scheme for organizing all the knowledge of humanity. Salvation lay in cutting up three-by-five cards, of which he had plenty.

Although Nelson was polite, charming, and smooth, I was too slow for his fast talk. But I got an *aha!* from his marvelous notion of hypertext. He was certain that every document in the world should be a footnote to some other document, and computers could make the links between them visible and permanent. This was a new idea at the time. But that was just the beginning. Scribbling on index cards, he sketched out complicated notions of transferring authorship back to creators and tracking payments as readers hopped along networks of documents, in what he called the "docuverse." He spoke of "transclusion" and "intertwingularity" as he described the grand utopian benefits of his embedded structure. It was going to save the world from stupidity!

I believed him. Despite his quirks, it was clear to me that a hyperlinked world was inevitable—someday. But as I look back now, after 30 years of living online, what surprises me about the genesis of the web is how much was missing from Vannevar Bush's vision, and even Nelson's docuverse, and especially my own expectations. We all missed the big story. Neither old ABC nor startup Yahoo! created the content for 5,000 web channels. Instead billions of users created the content for all the other users. There weren't 5,000 channels but 500 million channels, all customer generated. The disruption ABC could not imagine was that this "internet stuff" enabled the formerly dismissed passive consumers to become active creators. The revolution launched by the web was only marginally about hypertext and human knowledge. At its heart was a new kind of participation that has since developed into an emerging culture based on sharing. And the ways of "sharing" enabled by hyperlinks are now creating a new type of thinking—part human and part machine—found nowhere else on the planet or in history. The web has unleashed a new becoming.

Not only did we fail to imagine what the web would become, we still don't see it today. We are oblivious to the miracle it has blossomed into. Twenty years after its birth the immense scope of the web is hard to fathom. The total number of web pages, including those that are dynamically created upon request, exceeds 60 trillion. That's almost 10,000 pages per person alive. And this entire cornucopia has been created in less than 8,000 days.

The accretion of tiny marvels can numb us to the arrival of the stupendous. Today, from any window on the internet, you can get: an amazing variety of music and video, an evolving encyclopedia, weather forecasts, help-wanted ads, satellite images of any place on earth, up-to-the-minute news from around the globe, tax forms, TV guides, road maps with driving directions, real-time stock quotes, real estate listings with virtual walk-throughs and real-time prices, pictures of just about anything, latest sports scores, places to buy everything, records of political contributions, library catalogs, appliance manuals, live traffic reports, archives to major newspapers—all accessed instantly.

This view is spookily godlike. You can switch your gaze on a spot in the world from map to satellite to 3-D just by clicking. Recall the past? It's there. Or listen to the daily complaints and pleas of almost anyone who tweets or posts. (And doesn't everyone?) I doubt angels have a better view of humanity.

Why aren't we more amazed by this fullness? Kings of old would have gone to war to win such abilities. Only small children back then would have dreamed such a magic window could be real. I have reviewed the expectations of the wise experts from the 1980s, and I can affirm that this comprehensive wealth of material, available on demand and free of charge, was not in anyone's 20-year plan. At that time, anyone silly enough to trumpet the above list as a vision of the near future would have been confronted by the evidence: There wasn't enough money in all the investment firms in the entire world to fund such bounty. The success of the web at this scale was impossible.

But if we have learned anything in the past three decades, it is that the impossible is more plausible than it appears.

Nowhere in Ted Nelson's convoluted sketches of hypertext transclusion did the fantasy of a virtual flea market appear. Nelson hoped to franchise his Xanadu hypertext systems in the physical world at the scale of mom-and-pop cafés—you would go to a Xanadu store to do your hypertexting. Instead, the web erupted into open global flea markets like eBay, Craigslist, or Alibaba that handle several billion transactions every year and operate right into your bedroom. And here's the surprise: Users do most of the work—they photograph, they catalog, they post, and they market their own sales. And they police themselves; while the sites do call in the authorities to arrest serial abusers, the chief method of ensuring fairness is a system of user-generated ratings. Three billion feedback comments can work wonders.

What we all failed to see was how much of this brave new online world would be manufactured by users, not big institutions. The entirety of the content offered by Facebook, YouTube, Instagram, and Twitter is not created by their staff, but by their audience. Amazon's rise was a surprise not because it became an "everything store" (not hard to imagine), but because Amazon's customers (me and you) rushed to write the reviews that made the site's long-tail selection usable. Today, most major software producers have minimal help desks; their most enthusiastic customers advise and assist other customers on the company's support forum web pages, serving as high-quality customer support for new buyers. And in the greatest leverage of the common user, Google turns traffic and link patterns generated by 90 billion searches a month into the organizing intelligence for a new economy. This bottom-up overturning was also not in anyone's 20-year vision.

No web phenomenon has been more confounding than the infinite rabbit hole of YouTube and Facebook videos. Everything media experts knew about audiences—and they knew a lot—promoted the belief that audiences would never get off their butts and start making their own entertainment. The audience was a confirmed collective coach potato, as the ABC honchos assumed. Everyone knew writing and reading were dead; music was too much trouble to make when you could sit back and listen; video production was simply out of reach of amateurs in terms of cost and expertise. User-generated creations would never happen at a large scale, or if they happened they would not draw an audience, or if they drew an audience they would not matter. What a shock, then, to witness the near instantaneous rise of 50 million blogs in the early 2000s, with two new blogs appearing every second. And then a few years later the explosion of user-created videos—65,000 per day are posted to YouTube, or 300 video hours every minute, in 2015. And in recent years a ceaseless eruption of alerts, tips, and news headlines. Each user doing what ABC, AOL, *USA Today*—and almost everyone else—expected only ABC, AOL, *USA Today* would be doing. These user-created channels make no sense economically. Where are the time, energy, and resources coming from?

The audience.

The nutrition of participation nudges ordinary folks to invest huge hunks of energy and time into making free encyclopedias, creating free public tutorials for changing a flat tire, or cataloging the votes in the Senate. More and more of the web runs in this mode. One study a few years ago found that only 40 percent of the web is commercially manufactured. The rest is fueled by duty or passion.

Coming out of the industrial age, when mass-produced goods outperformed anything you could make yourself, this sudden tilt toward consumer involvement is a surprise. We thought, "That amateur do-it-yourself thing died long ago, back in the horse-and-buggy era." The enthusiasm for making things, for interacting more deeply than just choosing options, is the great force not reckoned—not seen—decades ago, even though it was already going on. This apparently primeval impulse for participation has upended the economy and is steadily turning the sphere of social networking—smart mobs, hive minds, and collaborative action—into the main event.

When a company opens part of its databases and functionality to users and other startups via a public API, or application programming interface, as Amazon, Google, eBay, Facebook, and most large platforms have, it is encouraging the participation of its users at new levels. People who take advantage of these capabilities are no longer a company's customers; they're the company's developers, vendors, laboratories, and marketers.

With the steady advance of new ways for customers and audiences to participate, the web has embedded itself into every activity and every region of the planet. Indeed, people's anxiety about the internet being out of the mainstream seems quaint now. The genuine 1990 worry about the internet being predominantly male was entirely misplaced. Everyone missed the party celebrating the 2002 flip point when women online first outnumbered men. Today, 51 percent of netizens are female. And, of course, the internet is not and has never been a teenage realm. In 2014 the average age of a user was roughly a bone-creaking 44 years old.

And what could be a better mark of universal acceptance than adoption by the Amish? I was visiting some Amish farmers recently. They fit the archetype perfectly: straw hats, scraggly beards, wives with bonnets, no electricity, no phones or TVs, horse and buggy outside. They have an undeserved reputation for resisting all technology, when actually they are just very late adopters. Still, I was amazed to hear them mention their websites.

"Amish websites?" I asked.

"For advertising our family business. We weld barbecue grills in our shop."

"Yes, but . . ."

"Oh, we use the internet terminal at the public library. And Yahoo!"

I knew then the takeover was complete. We are all becoming something new.

———

As we try to imagine this exuberant web three decades from now, our first impulse is to imagine it as Web 2.0—a better web. But the web in 2050 won't be a better web, just as the first version of the web was not better TV with more channels. It will have become something new, as different from the web today as the first web was from TV.

In a strict technical sense, the web today can be defined as the sum of all the things that you can google—that is, all files reachable with a hyperlink. Presently major portions of the digital world can't be googled. A lot of what happens in Facebook, or on a phone app, or inside a game world, or even inside a video can't be searched right now. In 30 years it will be. The tendrils of hyperlinks will keep expanding to connect all the bits. The events that take place in a console game will be as searchable as the news. You'll be able to look for things that occur inside a YouTube video. Say you want to find the exact moment on your phone when your sister received her acceptance to college. The web will reach this. It will also extend to physical objects, both manufactured and natural. A tiny, almost free chip embedded into products will connect them to the web and integrate their data. Most objects in your room will be connected, enabling you to google your room. Or google your house. We already have a hint of that. I can operate my thermostat and my music system from my phone. In three more decades, the rest of the world will overlap my devices. Unsurprisingly, the web will expand to the dimensions of the physical planet.

It will also expand in time. Today's web is remarkably ignorant of the past. It may supply you with a live webcam stream of Tahrir Square in Egypt, but accessing that square a year ago is nearly impossible. Viewing an earlier version of a typical website is not easy, but in 30 years we'll have time sliders enabling us to see any past version. Just as your phone's navigation directions through a city are improved by including previous days, weeks, and months of traffic patterns, so the web of 2050 will be informed by the context of the past. And the web will slide into the future as well.

From the moment you wake up, the web is trying to anticipate your intentions. Since your routines are noted, the web is attempting to get ahead of your actions, to deliver an answer almost before you ask a question. It is built to provide the files you need before the meeting, to suggest the perfect place to eat lunch with your friend, based on the weather, your location, what you ate this week, what you had the last time you met with your friend, and as many other factors as you might consider. You'll converse with the web. Rather than flick through stacks of friends' snapshots on your phone, you ask it about a friend. The web anticipates which photos you'd like to see and, depending on your reaction to those, may show you more or something from a different friend—or, if your next meeting is starting, the two emails you need to see. The web will more and more resemble a presence that you relate to rather than a place—the famous cyberspace of the 1980s—that you journey to. It will be a low-level constant presence like electricity: always around us, always on, and subterranean. By 2050 we'll come to think of the web as an ever-present type of conversation.

This enhanced conversation will unleash many new possibilities. Yet the digital world already feels bloated with too many choices and possibilities. There seem to be no slots for anything genuinely new in the next few years.

Can you imagine how awesome it would have been to be an ambitious entrepreneur back in 1985 at the dawn of the internet? At that time almost any dot-com name you desired was available. All you had to do was simply ask for the one you wanted. One-word domains, common names—they were all available. It didn't even cost anything to claim. This grand opportunity was true for years. In 1994 a *Wired* writer noticed that mcdonalds.com was still unclaimed, so with my encouragement he registered it. He then tried unsuccessfully to *give* it to McDonald's, but the company's cluelessness about the internet was so hilarious ("dot what?") that this tale became a famous story we published in *Wired*.

The internet was a wide-open frontier then. It was easy to be the first in any category you chose. Consumers had few expectations and the barriers were extremely low. Start a search engine! Be the first to open an online store! Serve up amateur videos! Of course, that was then. Looking back now, it seems as if waves of settlers have since bulldozed and developed every possible venue, leaving only the most difficult and gnarly specks for today's newcomers. Thirty years later the internet feels saturated with apps, platforms, devices, and more than enough content to demand our attention for the next million years. Even if you could manage to squeeze in another tiny innovation, who would notice it among our miraculous abundance?

But, but . . . here is the thing. In terms of the internet, nothing has happened yet! The internet is still at the beginning of its beginning. It is only becoming. If we could climb into a time machine, journey 30 years into the future, and from that vantage look back to today, we'd realize that most of the greatest products running the lives of citizens in 2050 were not invented until after 2016. People in the future will look at their holodecks and wearable virtual reality contact lenses and downloadable avatars and AI interfaces and say, "Oh, you didn't really have the internet"—or whatever they'll call it—"back then."

And they'd be right. Because from our perspective now, the greatest online things of the first half of this century are all before us. All these miraculous inventions are waiting for that crazy, no-one-told-me-it-was-impossible visionary to start grabbing the low-hanging fruit—the equivalent of the dot-com names of 1984.

Because here is the other thing the graybeards in 2050 will tell you: Can you imagine how awesome it would have been to be an innovator in 2016? It was a wide-open frontier! You could pick almost any category and add some AI to it, put it on the cloud. Few devices had more than one or two sensors in them, unlike the hundreds now. Expectations and barriers were low. It was easy to be the first. And then they would sigh. "Oh, if only we realized how possible everything was back then!"

So, the truth: Right now, today, in 2016 is the best time to start up. There has never been a better day in the whole history of the world to invent something. There has never been a better time with more opportunities, more openings, lower barriers, higher benefit/risk ratios, better returns, greater upside than now. Right now, this minute. This is the moment that folks in the future will look back at and say, "Oh, to have been alive and well back then!"

The last 30 years has created a marvelous starting point, a solid platform to build truly great things. But what's coming will be different, beyond, and other. The things we will make will be constantly, relentlessly becoming something else. And the coolest stuff of all has not been invented yet.

Today truly is a wide-open frontier. We are all becoming. It is the best time *ever* in human history to begin.

You are not late.

2

COGNIFYING

I t is hard to imagine anything that would "change everything" as much as cheap, powerful, ubiquitous artificial intelligence. To begin with, there's nothing as consequential as a dumb thing made smarter. Even a very tiny amount of useful intelligence embedded into an existing process boosts its effectiveness to a whole other level. The advantages gained from cognifying inert things would be hundreds of times more disruptive to our lives than the transformations gained by industrialization.

Ideally, this additional intelligence should be not just cheap, but free. A free AI, like the free commons of the web, would feed commerce and science like no other force we can imagine and would pay for itself in no time. Until recently, conventional wisdom held that supercomputers would be the first to host this artificial mind, and then perhaps we'd get mini minds at home, and then soon enough we'd add consumer models to the heads of our personal robots. Each AI would be a bounded entity. We would know where our thoughts ended and theirs began.

However, the first genuine AI will not be birthed in a stand-alone supercomputer, but in the superorganism of a billion computer chips known as the net. It will be planetary in dimensions, but thin, embedded, and loosely connected. It will be hard to tell where its thoughts begin and ours end. Any device that touches this networked AI will share—and contribute to—its intelligence. A lonely off-the-grid AI cannot learn as fast, or as smartly, as one that is plugged into 7 billion human minds, plus quintillions of online transistors, plus hundreds of exabytes of real-life data, plus the self-correcting feedback loops of the entire civilization. So the network itself will cognify into something that uncannily keeps getting better. Stand-alone synthetic minds are likely to be viewed as handicapped, a penalty one might pay in order to have AI mobility in distant places.

When this emerging AI arrives, its very ubiquity will hide it. We'll use its growing smartness for all kinds of humdrum chores, but it will be faceless, unseen. We will be able to reach this distributed intelligence in a million ways, through any digital screen anywhere on earth, so it will be hard to say where it is. And because this synthetic intelligence is a combination of human intelligence (all past human learning, all current humans online), it will be difficult to pinpoint exactly *what* it is as well. Is it our memory, or a consensual agreement? Are we searching it, or is it searching us?

The arrival of artificial thinking accelerates all the other disruptions I describe in this book; it is the ur-force in our future. We can say with certainty that cognification is inevitable, because it is already here.

———

Two years ago I made the trek to the sylvan campus of the IBM research labs in Yorktown Heights, New York, to catch an early glimpse of this rapidly appearing, long overdue arrival of artificial intelligence. This was the home of Watson, the electronic genius that conquered *Jeopardy!* in 2011. The original Watson is still here—it's about the size of a bedroom, with 10 upright refrigerator-shaped machines forming the four walls. The tiny interior cavity gives technicians access to the jumble of wires and cables on the machines' backs. It is surprisingly warm inside, as if the cluster were alive.

Today's Watson is very different. It no longer exists solely within a wall of cabinets but is spread across a cloud of open-standard servers that run several hundred "instances" of the AI at once. Like all things cloudy, Watson is served to simultaneous customers anywhere in the world, who can access it using their phones, their desktops, or their own data servers. This kind of AI can be scaled up or down on demand. Because AI improves as people use it, Watson is always getting smarter; anything it learns in one instance can be quickly transferred to the others. And instead of one single program, it's an aggregation of diverse software engines—its logic-deduction engine and its language-parsing engine might operate on different code, on different chips, in different locations— all cleverly integrated into a unified stream of intelligence.

Consumers can tap into that always-on intelligence directly, but also through third-party apps that harness the power of this AI cloud. Like many parents of a bright mind, IBM would like Watson to pursue a medical career, so it should come as no surprise that the primary application under development is a medical diagnosis tool. Most of the previous attempts to make a diagnostic AI have been pathetic failures, but Watson really works. When, in plain English, I give it the symptoms of a disease I once contracted in India, it gives me a list of hunches, ranked from most to least probable. The most likely cause, it declares, is giardia—the correct answer. This expertise isn't yet available to patients directly; IBM provides Watson's medical intelligence to partners like CVS, the retail pharmacy chain, helping it develop personalized health advice for customers with chronic diseases based on the data CVS collects. "I believe something like Watson will soon be the world's best diagnostician—whether machine or human," says Alan Greene, chief medical officer of Scanadu, a startup that is building a diagnostic device inspired by the *Star Trek* medical tricorder and powered by a medical AI. "At the rate AI technology is improving, a kid born today will rarely need to see a doctor to get a diagnosis by the time they are an adult."

Medicine is only the beginning. All the major cloud companies, plus dozens of startups, are in a mad rush to launch a Watson-like cognitive service. According to the analysis firm Quid, AI has attracted more than $18 billion in investments since 2009. In 2014 alone more than $2 billion was invested in 322 companies with AI-like technology. Facebook, Google, and their Chinese equivalents, TenCent and Baidu, have recruited researchers to join their in-house AI research teams. Yahoo!, Intel, Dropbox, LinkedIn, Pinterest, and Twitter have all purchased AI companies since 2014. Private investment in the AI sector has been expanding 70 percent a year on average for the past four years, a rate that is expected to continue.

One of the early stage AI companies Google purchased is DeepMind, based in London. In 2015 researchers at DeepMind published a paper in *Nature* describing how they taught an AI to learn to play 1980s-era arcade video games, like *Video Pinball*. They did not teach it how to play the games, but how *to learn to play* the games—a profound difference. They simply turned their cloud-based AI loose on an Atari game such as *Breakout*, a variant of *Pong*, and it learned on its own how to keep increasing its score. A video of the AI's progress is stunning. At first, the AI plays nearly randomly, but it gradually improves. After a half hour it misses only once every four times. By its 300th game, an hour into it, it never misses. It keeps learning so fast that in the second hour it figures out a loophole in the *Breakout* game that none of the millions of previous human players had discovered. This hack allowed it to win by tunneling around a wall in a way that even the game's creators had never imagined. At the end of several hours of first playing a game, with no coaching from the DeepMind creators, the algorithms, called deep reinforcement machine learning, could beat humans in half of the 49 Atari video games they mastered. AIs like this one are getting smarter every month, unlike human players.

Amid all this activity, a picture of our AI future is coming into view, and it is not the HAL 9000—a discrete machine animated by a charismatic (yet potentially homicidal) humanlike consciousness—or a Singularitan rapture of superintelligence. The AI on the horizon looks more like Amazon Web Services—cheap, reliable, industrial-grade digital smartness running behind everything, and almost invisible except when it blinks off. This common utility will serve you as much IQ as you want but no more than you need. You'll simply plug into the grid and get AI as if it was electricity. It will enliven inert objects, much as electricity did more than a century past. Three generations ago, many a tinkerer struck it rich by taking a tool and making an electric version. Take a manual pump; electrify it. Find a hand-wringer washer; electrify it. The entreprenuers didn't need to generate the electricity; they bought it from the grid and used it to automate the previously manual. Now everything that we formerly electrified we will cognify. There is almost nothing we can think of that cannot be made new, different, or more valuable by infusing it with some extra IQ. In fact, the business plans of the next 10,000 startups are easy to forecast: *Take X and add AI*. Find something that can be made better by adding online smartness to it.

An excellent example of the magic of adding AI to X can be seen in photography. In the 1970s I was a travel photographer hauling around a heavy bag of gear. In addition to a backpack with 500 rolls of film, I carried two brass Nikon bodies, a flash, and five extremely heavy glass lenses that weighed over a pound each. Photography needed "big glass" to capture photons in low light; it needed light-sealed cameras with intricate marvels of mechanical engineering to focus, measure, and bend light in thousandths of a second. What has happened since then? Today my point-and-shoot Nikon weighs almost nothing, shoots in almost no light, and can zoom from my nose to infinity. Of course, the camera in my phone is even tinier, always present, and capable of pictures as good as my old heavy clunkers. The new cameras are smaller, quicker, quieter, and cheaper not just because of advances in miniaturization, but because much of the traditional camera has been replaced by smartness. The X of photography has been cognified. Contemporary phone cameras eliminated the layers of heavy glass by adding algorithms, computation, and intelligence to do the work that physical lenses once did. They use the intangible smartness to substitute for a physical shutter. And the darkroom and film itself have been replaced by more computation and optical intelligence. There are even designs for a completely flat camera with no lens at all. Instead of any glass, a perfectly flat light sensor uses insane amounts of computational cognition to compute a picture from the different light rays falling on the unfocused sensor. Cognifying photography has revolutionized it because intelligence enables cameras to slip into anything (in a sunglass frame, in a color on clothes, in a pen) and do more, including calculate 3-D, HD, and many other options that earlier would have taken $100,000 and a van full of equipment to do. Now cognified photography is something almost any device can do as a side job.

A similar transformation is about to happen for every other X. Take chemistry, another physical endeavor requiring laboratories of glassware and bottles brimming with solutions. Moving atoms—what could be more physical? By adding AI to chemistry, scientists can perform virtual chemical experiments. They can smartly search through astronomical numbers of chemical combinations to reduce them to a few promising compounds worth examining in a lab. The X might be something low-tech, like interior design. Add utility AI to a system that matches levels of interest of clients as they walk through simulations of interiors. The design details are altered and tweaked by the pattern-finding AI based on customer response, then inserted back into new interiors for further testing. Through constant iterations, optimal personal designs emerge from the AI. You could also apply AI to law, using it to uncover evidence from mountains of paper to discern inconsistencies between cases, and then have it suggest lines of legal arguments.

The list of Xs is endless. The more unlikely the field, the more powerful adding AI will be. Cognified investments? Already happening with companies such as Betterment or Wealthfront. They add artificial intelligence to managed stock indexes in order to optimize tax strategies or balance holdings between portfolios. These are the kinds of things a professional money manager might do once a year, but the AI will do every day, or every hour.

Here are other unlikely realms waiting to be cognitively enhanced:

Cognified music—Music can be created in real time from algorithms, employed as the soundtrack for a video game or a virtual world. Depending on your actions, the music changes. Hundreds of hours of new personal music can be written by the AI for every player.

Cognified laundry—Clothes that tell the washing machines how they want to be washed. The wash cycle would adjust itself to the contents of each load as directed by the smart clothes.

Cognified marketing—The amount of attention an individual reader or watcher spends on an advertisement can be multiplied by their social influence (how many people followed them and what their influence was) in order to optimize attention and influence per dollar. Done at the scale of millions, this is a job for AI.

Cognified real estate—Matching buyers and sellers via an AI that can prompt "renters who liked this apartment also liked these . . ." It could then generate a financing package that worked for your particular circumstances.

Cognified nursing—Patients outfitted with sensors that track their bio markers 24 hours a day can generate highly personalized treatments that are adjusted and refined daily.

Cognified construction—Imagine project management software that is smart enough to take into account weather forecasts, port traffic delays, currency exchange rates, accidents, in addition to design changes.

Cognified ethics—Robo cars need to be taught priorities and behavior guidelines. The safety of pedestrians may precede the safety of drivers. Anything with some real autonomy that depends on code will also require smart ethical code as well.

Cognified toys—Toys more like pets. Furbies were primitive compared with the intense attraction that a smart petlike toy will invoke from children. Toys that can converse are lovable. Dolls may be the first really popular robots.

Cognified sports—Smart sensors and AI can create new ways to score and referee sporting games by tracking and interpreting subtle movements and collisions. Also, highly refined statistics can be extracted from every second of each athlete's activity to create elite fantasy sports leagues.

Cognified knitting—Who knows? But it will come!

Cognifying our world is a very big deal, and it's happening now.

———

Around 2002 I attended a private party for Google—before its IPO, when it was a small company focused only on search. I struck up a conversation with Larry Page, Google's brilliant cofounder. "Larry, I still don't get it. There are so many search companies. Web search, for free? Where does that get you?" My unimaginative blindness is solid evidence that predicting is hard, especially about the future, but in my defense this was before Google had ramped up its ad auction scheme to generate real income, long before YouTube or any other major acquisitions. I was not the only avid user of its search site who thought it would not last long. But Page's reply has always stuck with me: "Oh, we're really making an AI."

I've thought a lot about that conversation over the past few years as Google has bought 13 other AI and robotics companies in addition to DeepMind. At first glance, you might think that Google is beefing up its AI portfolio to improve its search capabilities, since search constitutes 80 percent of its revenue. But I think that's backward. Rather than use AI to make its search better, Google is using search to make its AI better. Every time you type a query, click on a search-generated link, or create a link on the web, you are training the Google AI. When you type "Easter Bunny" into the image search bar and then click on the most Easter Bunny–looking image, you are teaching the AI what an Easter Bunny looks like. Each of the 3 billion queries that Google conducts *each day* tutors the deep-learning AI over and over again. With another 10 years of steady improvements to its AI algorithms, plus a thousandfold more data and a hundred times more computing resources, Google will have an unrivaled AI. In a quarterly earnings conference call in the fall of 2015, Google CEO Sundar Pichai stated that AI was going to be "a core transformative way by which we are rethinking everything we are doing. . . . We are applying it across all our products, be it search, be it YouTube and Play, etc." My prediction: By 2026, Google's main product will not be search but AI.

This is the point where it is entirely appropriate to be skeptical. For almost 60 years, AI researchers have predicted that AI is right around the corner, yet until a few years ago it seemed as stuck in the future as ever. There was even a term coined to describe this era of meager results and even more meager research funding: the AI winter. Has anything really changed?

Yes. Three recent breakthroughs have unleashed the long-awaited arrival of artificial intelligence:

1. Cheap Parallel Computation

Thinking is an inherently parallel process. Billions of neurons in our brain fire simultaneously to create synchronous waves of computation. To build a neural network—the primary architecture of AI software—also requires many different processes to take place simultaneously. Each node of a neural network loosely imitates a neuron in the brain—mutually interacting with its neighbors to make sense of the signals it receives. To recognize a spoken word, a program must be able to hear all the phonemes in relation to one another; to identify an image, it needs to see every pixel in the context of the pixels around it—both deeply parallel tasks. But until recently, the typical computer processor could ping only one thing at a time.

That began to change more than a decade ago, when a new kind of chip, called a graphics processing unit, or GPU, was devised for the intensely visual—and parallel—demands of video games, in which millions of pixels in an image had to be recalculated many times a second. That required a specialized parallel computing chip, which was added as a supplement to the PC motherboard. The parallel graphics chips worked fantastically, and gaming soared in popularity. By 2005, GPUs were being produced in such quantities that they became so cheap they were basically a commodity. In 2009, Andrew Ng and a team at Stanford realized that GPU chips could run neural networks in parallel.

That discovery unlocked new possibilities for neural networks, which can include hundreds of millions of connections between their nodes. Traditional processors required several weeks to calculate all the cascading possibilities in a neural net with 100 million parameters. Ng found that a cluster of GPUs could accomplish the same thing in a day. Today neural nets running on GPUs are routinely used by cloud-enabled companies such as Facebook to identify your friends in photos or for Netflix to make reliable recommendations for its more than 50 million subscribers.

2. Big Data

Every intelligence has to be taught. A human brain, which is genetically primed to categorize things, still needs to see a dozen examples as a child before it can distinguish between cats and dogs. That's even more true for artificial minds. Even the best-programmed computer has to play at least a thousand games of chess before it gets good. Part of the AI breakthrough lies in the incredible avalanche of collected data about our world, which provides the schooling that AIs need. Massive databases, self-tracking, web cookies, online footprints, terabytes of storage, decades of search results, Wikipedia, and the entire digital universe became the teachers making AI smart. Andrew Ng explains it this way: "AI is akin to building a rocket ship. You need a huge engine and a lot of fuel. The rocket engine is the learning algorithms but the fuel is the huge amounts of data we can feed to these algorithms."

3. Better Algorithms

Digital neural nets were invented in the 1950s, but it took decades for computer scientists to learn how to tame the astronomically huge combinatorial relationships between a million—or a hundred million—neurons. The key was to organize neural nets into stacked layers. Take the relatively simple task of recognizing that a face is a face. When a group of bits in a neural net is found to trigger a pattern—the image of an eye, for instance—that result ("It's an eye!") is moved up to another level in the neural net for further parsing. The next level might group two eyes together and pass that meaningful chunk on to another level of hierarchical structure that associates it with the pattern of a nose. It can take many millions of these nodes (each one producing a calculation feeding others around it), stacked up to 15 levels high, to recognize a human face. In 2006, Geoff Hinton, then at the University of Toronto, made a key tweak to this method, which he dubbed "deep learning." He was able to mathematically optimize results from each layer so that the learning accumulated faster as it proceeded up the stack of layers. Deep-learning algorithms accelerated enormously a few years later when they were ported to GPUs. The code of deep learning alone is insufficient to generate complex logical thinking, but it is an essential component of all current AIs, including IBM's Watson; DeepMind, Google's search engine; and Facebook's algorithms.

This perfect storm of cheap parallel computation, bigger data, and deeper algorithms generated the 60-years-in-the-making overnight success of AI. And this convergence suggests that as long as these technological trends continue—and there's no reason to think they won't—AI will keep improving.

As it does, this cloud-based AI will become an increasingly ingrained part of our everyday life. But it will come at a price. Cloud computing empowers the law of increasing returns, sometimes called the network effect, which holds that the value of a network increases much faster as it grows bigger. The bigger the network, the more attractive it is to new users, which makes it even bigger and thus more attractive, and so on. A cloud that serves AI will obey the same law. The more people who use an AI, the smarter it gets. The smarter it gets, the more people who use it. The more people who use it, the smarter it gets. And so on. Once a company enters this virtuous cycle, it tends to grow so big so fast that it overwhelms any upstart competitors. As a result, our AI future is likely to be ruled by an oligarchy of two or three large, general-purpose cloud-based commercial intelligences.

In 1997, Watson's precursor, IBM's Deep Blue, beat the reigning chess grand master Garry Kasparov in a famous man-versus-machine match. After machines repeated their victories in a few more matches, humans largely lost interest in such contests. You might think that was the end of the story (if not the end of human history), but Kasparov realized that he could have performed better against Deep Blue if he'd had the same instant access to a massive database of all previous chess moves that Deep Blue had. If this database tool was fair for an AI, why not for a human? Let the human mastermind be augmented by a database just as Deep Blue's was. To pursue this idea, Kasparov pioneered the concept of man-plus-machine matches, in which AI augments human chess players rather than competes against them.

Now called freestyle chess matches, these are like mixed martial arts fights, where players use whatever combat techniques they want. You can play as your unassisted human self, or you can act as the hand for your supersmart chess computer, merely moving its board pieces, or you can play as a "centaur," which is the human/AI cyborg that Kasparov advocated. A centaur player will listen to the moves suggested by the AI but will occasionally override them—much the way we use the GPS navigation intelligence in our cars. In the championship Freestyle Battle 2014, open to all modes of players, pure chess AI engines won 42 games, but centaurs won 53 games. Today the best chess player alive is a centaur. It goes by the name of Intagrand, a team of several humans and several different chess programs.

But here's the even more surprising part: The advent of AI didn't diminish the performance of purely human chess players. Quite the opposite. Cheap, supersmart chess programs inspired more people than ever to play chess, at more tournaments than ever, and the players got better than ever. There are more than twice as many grand masters now as there were when Deep Blue first beat Kasparov. The top-ranked human chess player today, Magnus Carlsen, trained with AIs and has been deemed the most computerlike of all human chess players. He also has the highest human grand master rating of all time.

If AI can help humans become better chess players, it stands to reason that it can help us become better pilots, better doctors, better judges, better teachers.

Yet most of the commercial work completed by AI will be done by nonhuman-like programs. The bulk of AI will be special purpose software brains that can, for example, translate any language into any other language, but do little else. Drive a car, but not converse. Or recall every pixel of every video on YouTube, but not anticipate your work routines. In the next 10 years, 99 percent of the artificial intelligence that you will interact with, directly or indirectly, will be nerdly narrow, supersmart specialists.

In fact, robust intelligence may be a liability—especially if by "intelligence" we mean our peculiar self-awareness, all our frantic loops of introspection and messy currents of self-consciousness. We want our self-driving car to be inhumanly focused on the road, not obsessing over an argument it had with the garage. The synthetic Dr. Watson at our hospital should be maniacal in its work, never wondering whether it should have majored in finance instead. What we want instead of conscious intelligence is artificial smartness. As AIs develop, we might have to engineer ways to *prevent* consciousness in them. Our most premium AI services will likely be advertised as consciousness-free.

Nonhuman intelligence is not a bug; it's a feature. The most important thing to know about thinking machines is that they will think different.

Because of a quirk in our evolutionary history, we are cruising as the only self-conscious species on our planet, leaving us with the incorrect idea that human intelligence is singular. It is not. Our intelligence is a society of intelligences, and this suite occupies only a small corner of the many types of intelligences and consciousnesses that are possible in the universe. We like to call our human intelligence "general purpose," because compared with other kinds of minds we have met, it can solve more types of problems, but as we build more and more synthetic minds we'll come to realize that human thinking is not general at all. It is only one species of thinking.

The kind of thinking done by the emerging AIs today is not like human thinking. While they can accomplish tasks—such as playing chess, driving a car, describing the contents of a photograph—that we once believed only humans could do, they don't do it in a humanlike fashion. I recently uploaded 130,000 of my personal snapshots—my entire archive—to Google Photo, and the new Google AI remembers all the objects in all the images from my life. When I ask it to show me any image with a bicycle in it, or a bridge, or my mother, it will instantly display them. Facebook has the ability to ramp up an AI that can view a photo portrait of any person on earth and correctly identify them out of some 3 billion people online. Human brains cannot scale to this degree, which makes this artificial ability very *unhuman*. We are notoriously bad at statistical thinking, so we are making intelligences with very good statistical skills, in order that they don't think like us. One of the advantages of having AIs drive our cars is that they *won't* drive like humans, with our easily distracted minds.

In a superconnected world, thinking different is the source of innovation and wealth. Just being smart is not enough. Commercial incentives will make industrial-strength AI ubiquitous, embedding cheap smartness into all that we make. But a bigger payoff will come when we start inventing new kinds of intelligences and entirely new ways of thinking—in the way a calculator is a genius in arithmetic. Calculation is only one type of smartness. We don't know what the full taxonomy of intelligence is right now. Some traits of human thinking will be common (as common as bilateral symmetry, segmentation, and tubular guts are in biology), but the possibility space of viable minds will likely contain traits far outside what we have evolved. It is not necessary that this type of thinking be faster than humans', greater, or deeper. In some cases it will be simpler.

The variety of potential minds in the universe is vast. Recently we've begun to explore the species of animal minds on earth, and as we do we have discovered, with increasing respect, that we have met many other kinds of intelligences already. Whales and dolphins keep surprising us with their intricate and weirdly different intelligence. Precisely how a mind can be different or superior to our minds is very difficult to imagine. One way that would help us to imagine what greater yet different intelligences would be like is to begin to create a taxonomy of the variety of minds. This matrix of minds would include animal minds, and machine minds, and possible minds, particularly transhuman minds, like the ones that science fiction writers have come up with.

The reason this fanciful exercise is worth doing is because, while it is inevitable that we will manufacture intelligences in all that we make, it is not inevitable or obvious what their character will be. Their character will dictate their economic value and their roles in our culture. Outlining the possible ways that a machine might be smarter than us (even in theory) will assist us in both directing this advance and managing it. A few really smart people, like astronomer Stephen Hawking and genius inventor Elon Musk, worry that making supersmart AIs could be our last invention before they replace us (though I don't believe this), so exploring possible types is prudent.

Imagine we land on an alien planet. How would we measure the level of the intelligences we encounter there? This is an extremely difficult question because we have no real definition of our own intelligence, in part because until now we didn't need one.

In the real world—even in the space of powerful minds—trade-offs rule. One mind cannot do all mindful things perfectly well. A particular species of mind will be better in certain dimensions, but at a cost of lesser abilities in other dimensions. The smartness that guides a self-driving truck will be a different species than the one that evaluates mortgages. The AI that will diagnose your illness will be significantly different from the artificial smartness that oversees your house. The superbrain that predicts the weather accurately will be in a completely different kingdom of mind from the intelligence woven into your clothes. The taxonomy of minds must reflect the different ways in which minds are engineered with these trade-offs. In the short list below I include only those kinds of minds that we might consider superior to us; I've omitted the thousands of species of mild machine smartness—like the brains in a calculator—that will cognify the bulk of the internet of things.

Some possible new minds:

- A mind like a human mind, just faster in answering (the easiest AI mind to imagine).
- A very slow mind, composed primarily of vast storage and memory.
- A global supermind composed of millions of individual dumb minds in concert.
- A hive mind made of many very smart minds, but unaware it/they are a hive.
- A borg supermind composed of many smart minds that are very aware they form a unity.
- A mind trained and dedicated to enhancing your personal mind, but useless to anyone else.
- A mind capable of imagining a greater mind, but incapable of making it.
- A mind capable of creating a greater mind, but not self-aware enough to imagine it.
- A mind capable of successfully making a greater mind, once.
- A mind capable of creating a greater mind that can create a yet greater mind, etc.

- A mind with operational access to its source code, so it can routinely mess with its own processes.
- A superlogic mind without emotion.
- A general problem-solving mind, but without any self-awareness.
- A self-aware mind, but without general problem solving.
- A mind that takes a long time to develop and requires a protector mind until it matures.
- An ultraslow mind spread over large physical distance that appears "invisible" to fast minds.
- A mind capable of cloning itself exactly many times quickly.
- A mind capable of cloning itself and remaining in unity with its clones.
- A mind capable of immortality by migrating from platform to platform.
- A rapid, dynamic mind capable of changing the process and character of its cognition.
- A nanomind that is the smallest possible (size and energy profile) self-aware mind.
- A mind specializing in scenario and prediction making.
- A mind that never erases or forgets anything, including incorrect or false information.
- A half-machine, half-animal symbiont mind.
- A half-machine, half-human cyborg mind.
- A mind using quantum computing whose logic is not understandable to us.

————

If any of these imaginary minds are possible, it will be in the future beyond the next two decades. The point of this speculative list is to emphasize that all cognition is specialized. The types of artificial minds we are making now and will make in the coming century will be designed to perform specialized tasks, and usually tasks that are beyond what we can do. Our most important mechanical inventions are not machines that do what humans do better, but machines that can do things we can't do at all. Our most important thinking machines will not be machines that can think what we think faster, better, but those that think what we can't think.

To really solve the current grand mysteries of quantum gravity, dark energy, and dark matter, we'll probably need other intelligences beside human. And the extremely complex harder questions that will come after those hard questions may require even more distant and complex intelligences. Indeed, we may need to invent intermediate intelligences that can help us design yet more rarefied intelligences that we could not design alone. We need ways to think different.

Today, many scientific discoveries require hundreds of human minds to solve, but in the near future there may be classes of problems so deep that they require hundreds of different *species* of minds to solve. This will take us to a cultural edge because it won't be easy to accept the answers from an alien intelligence. We already see that reluctance in our difficulty in approving mathematical proofs done by computer. Some mathematical proofs have become so complex only computers are able to rigorously check every step, but these proofs are not accepted as "proof" by all mathematicians. The proofs are not understandable by humans alone so it is necessary to trust a cascade of algorithms, and this demands new skills in knowing when to trust these creations. Dealing with alien intelligences will require similar skills, and a further broadening of ourselves. An embedded AI will change how we do science. Really intelligent instruments will speed and alter our measurements; really huge sets of constant real-time data will speed and alter our model making; really smart documents will speed and alter our acceptance of when we "know" something. The scientific method is a way of knowing, but it has been based on how humans know. Once we add a new kind of intelligence into this method, science will have to know, and progress, according to the criteria of new minds. At that point everything changes.

AI could just as well stand for "alien intelligence." We have no certainty we'll contact extraterrestrial beings from one of the billion earth-like planets in the sky in the next 200 years, but we have almost 100 percent certainty that we'll manufacture an alien intelligence by then. When we face these synthetic aliens, we'll encounter the same benefits and challenges that we expect from contact with ET. They will force us to reevaluate our roles, our beliefs, our goals, our identity. What are humans for? I believe our first answer will be: Humans are for inventing new kinds of intelligences that biology could not evolve. Our job is to make machines that think different—to create alien intelligences. We should really call AIs "AAs," for "artificial aliens."

An AI will think about science like an alien, vastly different than any human scientist, thereby provoking us humans to think about science differently. Or to think about manufacturing materials differently. Or clothes. Or financial derivatives. Or any branch of science or art. The alienness of artificial intelligence will become more valuable to us than its speed or power.

Artificial intelligence will help us better understand what we mean by intelligence in the first place. In the past, we would have said *only* a superintelligent AI could drive a car or beat a human at *Jeopardy!* or recognize a billion faces. But once our computers did each of those things in the last few years, we considered that achievement obviously mechanical and hardly worth the label of true intelligence. We label it "machine learning." Every achievement in AI redefines that success as "not AI."

But we haven't just been redefining what we mean by AI—we've been redefining what it means to be human. Over the past 60 years, as mechanical processes have replicated behaviors and talents we thought were unique to humans, we've had to change our minds about what sets us apart. As we invent more species of AI, we will be forced to surrender more of what is supposedly unique about humans. Each step of surrender— we are not the only mind that can play chess, fly a plane, make music, or invent a mathematical law—will be painful and sad. We'll spend the next three decades—indeed, perhaps the next century—in a permanent identity crisis, continually asking ourselves what humans are good for. If we aren't unique toolmakers, or artists, or moral ethicists, then what, if anything, makes us special? In the grandest irony of all, the greatest benefit of an everyday, utilitarian AI will not be increased productivity or an economics of abundance or a new way of doing science—although all those will happen. The greatest benefit of the arrival of artificial intelligence is that AIs will help define humanity. We need AIs to tell us who we are.

———

The alien minds that we'll pay the most attention to in the next few years are the ones we give bodies to. We call them robots. They too will come in all shapes, sizes, and configurations—manifesting in diverse species, so to speak. Some will roam like animals, but many will be immobile like plants or diffuse like a coral reef. Robots are already here, quietly. Very soon louder, smarter ones are inevitable. The disruption they cause will touch our core.

Imagine that seven out of ten working Americans got fired tomorrow. What would they all do?

It's hard to believe you'd have an economy at all if you gave pink slips to more than half the labor force. But that—in slow motion—is what the industrial revolution did to the workforce of the early 19th century. Two hundred years ago, 70 percent of American workers lived on the farm. Today automation has eliminated all but 1 percent of their jobs, replacing them (and their work animals) with machines. But the displaced workers did not sit idle. Instead, automation created hundreds of millions of jobs in entirely new fields. Those who once farmed were now manning the legions of factories that churned out farm equipment, cars, and other industrial products. Since then, wave upon wave of new occupations have arrived—appliance repair person, offset printer, food chemist, photographer, web designer—each building on previous automation. Today, the vast majority of us are doing jobs that no farmer from the 1800s could have imagined.

It may be hard to believe, but before the end of this century, 70 percent of today's occupations will likewise be replaced by automation—including the job you hold. In other words, robots are inevitable and job replacement is just a matter of time. This upheaval is being led by a second wave of automation, one that is centered on artificial cognition, cheap sensors, machine learning, and distributed smarts. This broad automation will touch all jobs, from manual labor to knowledge work.

First, machines will consolidate their gains in already automated industries. After robots finish replacing assembly line workers, they will replace the workers in warehouses. Speedy bots able to lift 150 pounds all day long will retrieve boxes, sort them, and load them onto trucks. Robots like this already work in Amazon's warehouses. Fruit and vegetable picking will continue to be robotized until no humans pick outside of specialty farms. Pharmacies will feature a single pill-dispensing robot in the back while the pharmacists focus on patient consulting. In fact, prototype pill-dispensing robots are already up and running in hospitals in California. To date, they have not messed up a single prescription, something that cannot be said of any human pharmacist. Next, the more dexterous chores of cleaning in offices and schools will be taken over by late-night robots, starting with easy-to-do floors and windows and eventually advancing to toilets. The highway parts of long-haul trucking routes will be driven by robots embedded in truck cabs. By 2050 most truck drivers won't be human. Since truck driving is currently the most common occupation in the U.S., this is a big deal.

All the while, robots will continue their migration into white-collar work. We already have artificial intelligence in many of our machines; we just don't call it that. Witness one of Google's newest computers that can write an accurate caption for any photo it is given. Pick a random photo from the web, and the computer will "look" at it, then caption it perfectly. It can keep correctly describing what's going on in a series of photos as well as a human, but never tire. Google's translation AI turns a phone into a personal translator. Speak English into the microphone and it immediately repeats what you said in understandable Chinese, or Russian, or Arabic, or dozens of other languages. Point the phone to the recipient and the app will instantly translate their reply. The machine translator does Turkish to Hindi, or French to Korean, etc. It can of course translate any text. High-level diplomatic translators won't lose their jobs for a while, but day-to-day translating chores in business will all be better done by machines. In fact, any job dealing with reams of paperwork will be taken over by bots, including much of medicine. The rote tasks of any information-intensive job can be automated. It doesn't matter if you are a doctor, translator, editor, lawyer, architect, reporter, or even programmer: The robot takeover will be epic.

We are already at the inflection point.

We have preconceptions about how an intelligent robot should look and act, and these can blind us to what is already happening around us. To demand that artificial intelligence be humanlike is the same flawed logic as demanding that artificial flying be birdlike, with flapping wings. Robots, too, will think different.

Consider Baxter, a revolutionary new workbot from Rethink Robotics. Designed by Rodney Brooks, the former MIT professor who invented the bestselling Roomba vacuum cleaner and its descendants, Baxter is an early example of a new class of industrial robots created to work alongside humans. Baxter does not look impressive. Sure, it's got big strong arms and a flat-screen display like many industrial bots. And Baxter's hands perform repetitive manual tasks, just as factory robots do. But it's different in three significant ways.

First, it can look around and indicate where it is looking by shifting the cartoon eyes on its head. It can perceive humans working near it and avoid injuring them. And workers can see whether it sees them. Previous industrial robots couldn't do this, which meant that working robots had to be physically segregated from humans. The typical factory robot today is imprisoned within a chain-link fence or caged in a glass case. They are simply too dangerous to be around, because they are oblivious to others. This isolation prevents such robots from working in a small shop, where isolation is not practical. Optimally, workers should be able to get materials to and from the robot or to tweak its controls by hand throughout the workday; isolation makes that difficult. Baxter, however, is aware. Using force-feedback technology to feel if it is colliding with a person or another bot, it is courteous. You can plug it into a wall socket in your garage and easily work right next to it.

Second, anyone can train Baxter. It is not as fast, strong, or precise as other industrial robots, but it is smarter. To train the bot, you simply grab its arms and guide them in the correct motions and sequence. It's a kind of "watch me do this" routine. Baxter learns the procedure and then repeats it. Any worker is capable of this show and tell; you don't even have to be literate. Previous workbots required highly educated engineers and crack programmers to write thousands of lines of code (and then debug them) in order to instruct the robot in the simplest change of task. The code has to be loaded in batch mode—i.e., in large, infrequent batches—because the robot cannot be reprogrammed while it is being used. Turns out the real cost of the typical industrial robot is not its hardware but its operation. Industrial robots cost $100,000-plus to purchase but can require four times that amount over a lifespan to program, train, and maintain. The costs pile up until the average lifetime bill for an industrial robot is half a million dollars or more.

The third difference, then, is that Baxter is cheap. Priced at $25,000, it's in a different league compared with the $500,000 total bill of its predecessors. It is as if those established robots, with their batch-mode programming, are the mainframe computers of the robot world and Baxter is the first PC robot. It is likely to be dismissed as a hobbyist toy, missing key features like sub-millimeter precision. But as with the PC and unlike the ancient mainframe, the user can interact with it directly, immediately, without waiting for experts to mediate—and use it for nonserious, even frivolous things. It's cheap enough that small-time manufacturers can afford one to package up their wares or custom paint their product or run their 3-D printing machine. Or you could staff up a factory that makes iPhones.

Baxter was invented in a century-old brick building near the Charles River in Boston. In 1895 the building was a manufacturing marvel in the very center of the new manufacturing world. It even generated its own electricity. For a hundred years the factories inside its walls changed the world around us. Now the capabilities of Baxter and the approaching cascade of superior robot workers spur inventor Brooks to speculate on how these robots will shift manufacturing in a disruption greater than the last revolution. Looking out his office window at the former industrial neighborhood, he says, "Right now we think of manufacturing as happening in China. But as manufacturing costs sink because of robots, the costs of transportation become a far greater factor than the cost of production. Nearby will be cheap. So we'll get this network of locally franchised factories, where most things will be made within five miles of where they are needed."

That may be true for making stuff, but a lot of remaining jobs for humans are service jobs. I ask Brooks to walk with me through a local

McDonald's and point out the jobs that his kind of robots can replace. He demurs and suggests it might be 30 years before robots will cook for us. "In a fast-food place you're not doing the same task very long. You're always changing things on the fly, so you need special solutions. We are not trying to sell a specific solution. We are building a general-purpose machine that other workers can set up themselves and work alongside." And once we can cowork with robots right next to us, it's inevitable that our tasks will bleed together, and soon our old work will become theirs— and our new work will become something we can hardly imagine.

To understand how robot replacement will happen, it's useful to break down our relationship with robots into four categories.

1. Jobs Humans Can Do but Robots Can Do Even Better

Humans can weave cotton cloth with great effort, but automated looms make perfect cloth by the mile for a few cents. The only reason to buy handmade cloth today is because you want the imperfections humans introduce. There's very little reason to want an imperfect car. We no longer value irregularities while traveling 70 miles per hour on a highway— so we figure that the fewer humans touching our car as it is being made, the better.

And yet for more complicated chores, we still tend to mistakenly believe computers and robots can't be trusted. That's why we've been slow to acknowledge how they've mastered some conceptual routines, in certain cases even surpassing their mastery of physical routines. A computerized brain known as autopilot can fly a 787 jet unaided for all but seven minutes of a typical flight. We place human pilots in the cockpit to fly those seven minutes and for "just in case" insurance, but the needed human pilot time is decreasing rapidly. In the 1990s, computerized mortgage appraisals replaced human appraisers wholesale. Much tax preparation has gone to computers, as well as routine X-ray analysis and pretrial evidence gathering—all once done by highly paid smart people. We've accepted utter reliability in robot manufacturing; soon we'll accept the fact that robots can do it better in services and knowledge work too.

2. Jobs Humans Can't Do but Robots Can

A trivial example: Humans have trouble making a single brass screw unassisted, but automation can produce a thousand exact ones per hour. Without automation, we could not make a single computer chip—a job that requires degrees of precision, control, and unwavering attention that our animal bodies don't possess. Likewise no human—indeed no group of humans, no matter their education—can quickly search through all the web pages in the world to uncover the one page revealing the price of eggs in Kathmandu yesterday. Every time you click on the search button you are employing a robot to do something we as a species are unable to do alone.

While the displacement of formerly human jobs gets all the headlines, the greatest benefits bestowed by robots and automation come from their occupation of jobs we are unable to do. We don't have the attention span to inspect every square millimeter of every CAT scan looking for cancer cells. We don't have the millisecond reflexes needed to inflate molten glass into the shape of a bottle. We don't have an infallible memory to keep track of every pitch in Major League baseball and calculate the probability of the next pitch in real time.

We aren't giving "good jobs" to robots. Most of the time we are giving them jobs we could never do. Without them, these jobs would remain undone.

3. Jobs We Didn't Know We Wanted Done

This is the greatest genius of the robot takeover: With the assistance of robots and computerized intelligence, we already can do things we never imagined doing 150 years ago. We can today remove a tumor in our gut through our navel, make a talking-picture video of our wedding, drive a cart on Mars, print a pattern on fabric that a friend mailed to us as a message through the air. We are doing, and are sometimes paid for doing, a million new activities that would have dazzled and shocked the farmers of 1800. These new accomplishments are not merely chores that were difficult before. Rather they are dreams created chiefly by the capabilities of the machines that can do them. They are jobs the machines make up.

Before we invented automobiles, air-conditioning, flat-screen video displays, and animated cartoons, no one living in ancient Rome wished they could watch pictures move while riding to Athens in climate-controlled comfort. I did that recently. One hundred years ago not a single citizen of China would have told you that they would rather buy a tiny glassy slab that allowed them to talk to faraway friends before they would buy indoor plumbing. But every day peasant farmers in China without plumbing purchase smartphones. Crafty AIs embedded in first-person shooter games have given millions of teenage boys the urge, the need, to become professional game designers—a dream that no boy in Victorian times ever had. In a very real way our inventions assign us our jobs. Each successful bit of automation generates new occupations—occupations we would not have fantasized about without the prompting of the automation.

To reiterate, the bulk of new tasks created by automation are tasks only other automation can handle. Now that we have search engines like Google, we set the servant upon a thousand new errands. Google, can you tell me where my phone is? Google, can you match the people suffering depression with the doctors selling pills? Google, can you predict when the next viral epidemic will erupt? Technology is indiscriminate this way, piling up possibilities and options for both humans and machines.

It is a safe bet that the highest-earning professions in the year 2050 will depend on automations and machines that have not been invented yet. That is, we can't see these jobs from here, because we can't yet see the machines and technologies that will make them possible. Robots create jobs that we did not even know we wanted done.

4. Jobs Only Humans Can Do—at First

The one thing humans can do that robots can't (at least for a long while) is to decide what it is that humans want to do. This is not a trivial semantic trick; our desires are inspired by our previous inventions, making this a circular question.

When robots and automation do our most basic work, making it relatively easy for us to be fed, clothed, and sheltered, then we are free to ask, "What are humans for?" Industrialization did more than just extend the average human lifespan. It led a greater percentage of the population to decide that humans were meant to be ballerinas, full-time musicians, mathematicians, athletes, fashion designers, yoga masters, fan-fiction authors, and folks with one-of-a-kind titles on their business cards. With the help of our machines, we could take up these roles—but, of course, over time the machines will do these as well. We'll then be empowered to dream up yet more answers to the question "What should we do?" It will be many generations before a robot can answer that.

This postindustrial economy will keep expanding because each person's task (in part) will be to invent new things to do that will later become repetitive jobs for the robots. In the coming years robot-driven cars and trucks will become ubiquitous; this automation will spawn the new human occupation for former truck drivers of trip optimizer, a person who tweaks the traffic algorithms for optimal energy and time usage. Routine robosurgery will necessitate the new medical skills of keeping complex machines sterile. When automatic self-tracking of all your activities becomes the normal thing to do, a new breed of professional analysts will arise to help you make sense of the data. And of course we will need a whole army of robot nannies, dedicated to keeping your personal robots up and running. Each of these new vocations will in turn be taken over by automation later.

The real revolution erupts when everyone has personal workbots, the descendants of Baxter, at their beck and call. Imagine you are one of the 0.1 percent of people who still farm. You run a small organic farm with direct sales to your customers. You still have a job as a farmer, but robots do most of the actual farmwork. Your fleets of worker bots do all the outside work under the hot sun—weeding, pest control, and harvesting of produce—as directed by a very smart mesh of probes in the soil. Your new job as farmer is overseeing the farming system. One day your task might be to research which variety of heirloom tomato to plant; the next day to find out what your customers crave; the following day might be the time to update the information on your custom labels. The bots perform everything else that can be measured.

Right now it seems unthinkable: We can't imagine a bot that can assemble a stack of ingredients into a gift or manufacture spare parts for our lawn mower or fabricate materials for our new kitchen. We can't imagine our nephews and nieces running a dozen workbots in their garage, churning out inverters for their friend's electric vehicle startup. We can't imagine our children becoming appliance designers, making custom batches of liquid nitrogen dessert machines to sell to the millionaires in China. But that's what personal robot automation will enable.

Everyone will have access to a personal robot, but simply owning one will not guarantee success. Rather, success will go to those who best optimize the process of working with bots and machines. Geographical clusters of production will matter, not for any differential in labor costs but because of the differential in human expertise. It's human-robot symbiosis. Our human assignment will be to keep making jobs for robots— and that is a task that will never be finished. So we will always have at least that one "job."

———

In the coming years, our relationships with robots will become ever more complex. But already a recurring pattern is emerging. No matter what your current job or your salary, you will progress through a predictable cycle of denial again and again. Here are the Seven Stages of Robot Replacement:

1. A robot/computer cannot possibly do the tasks I do.
2. *[Later.]*
   OK, it can do a lot of those tasks, but it can't do everything I do.
3. *[Later.]*
   OK, it can do everything I do, except it needs me when it breaks down, which is often.
4. *[Later.]*
   OK, it operates flawlessly on routine stuff, but I need to train it for new tasks.
5. *[Later.]*
   OK, OK, it can have my old boring job, because it's obvious that was not a job that humans were meant to do.
6. *[Later.]*
   Wow, now that robots are doing my old job, my new job is much more interesting and pays more!
7. *[Later.]*
   I am so glad a robot/computer cannot possibly do what I do now. *[Repeat.]*

This is not a race against the machines. If we race against them, we lose. This is a race *with* the machines. You'll be paid in the future based on how well you work with robots. Ninety percent of your coworkers will be unseen machines. Most of what you do will not be possible without them. And there will be a blurry line between what you do and what they do. You might no longer think of it as a job, at least at first, because anything that resembles drudgery will be handed over to robots by the accountants.

We need to let robots take over. Many of the jobs that politicians are fighting to keep away from robots are jobs that no one wakes up in the morning really wanting to do. Robots will do jobs we have been doing, and do them much better than we can. They will do jobs we can't do at all. They will do jobs we never imagined even needed to be done. And they will help us discover new jobs for ourselves, new tasks that expand who we are. They will let us focus on becoming more human than we were.

It is inevitable. Let the robots take our jobs, and let them help us dream up new work that matters.

3

FLOWING

The internet is the world's largest copy machine. At its most fundamental level this machine copies every action, every character, every thought we make while we ride upon it. In order to send a message from one corner of the internet to another, the protocols of communication demand that the whole message be copied along the way several times. Some bits of data may be copied dozens of times in an ordinary day as they cycle through memory, cache, server, routers, and back. Tech companies make a lot of money selling equipment that facilitates this ceaseless copying. If something can be copied—a song, a movie, a book—and it touches the internet, it will be copied.

The digital economy runs on this river of freely flowing copies. In fact, our digital communication network has been engineered so that copies flow with as little friction as possible. Copies flow so freely we could think of the internet as a superconductor, where once a copy is introduced it will continue to flow through the network forever, much like electricity in a superconductive wire. This is what it means when something goes viral. The copies are recopied, and those duplications ripple outward launching new copies, in an endless contagious wave. Once a copy has touched the internet, it never leaves.

This superdistribution system has become the foundation of our economy and wealth. The instant reduplication of data, ideas, and media underpins the major sectors of a 21st-century economy. Copy-prone products, such as software, music, movies, and games are among the most valuable exports of the U.S., and they issue from the industries where the U.S. has a globally competitive advantage. American wealth therefore sits upon a very large device that copies promiscuously and constantly.

We can't stop massive indiscriminate copying. Not only would that sabotage the engine of wealth if we could, but it would halt the internet itself. Free-flowing copies are baked into the nature of this global communications system. The technology of the net needs to copy without constraint. The flow of copies is inevitable.

Our civilization's previous economy was built upon warehouses of fixed goods and factories stockpiled with solid cargo. These physical stocks are still necessary, but they are no longer sufficient for wealth and happiness. Our attention has moved away from stocks of solid goods to flows of intangibles, like copies. We value not only the atoms in a thing, but their immaterial arrangement and design and, even more, their ability to adapt and flow in response to our needs.

Formerly solid products made of steel and leather are now sold as fluid services that keep updating. Your solid car parked in a driveway has been transformed into a personal on-demand transportation service supplied by Uber, Lyft, Zip, and Sidecar—which are improving faster than automobiles are. Grocery shopping is no longer a hit-or-miss affair; now a steady flow of household replenishables streams into our homes uninterrupted. You get a better telephone every few months because a flow of new operating systems install themselves on your smartphone, adding new features and new benefits that in the past would have required new hardware. Then, when you do get new hardware, the service maintains the familiar operating system you had, flowing your personalization onto the new device. This total sequence of perpetual upgrades is continuous. It's a dream come true for our insatiable human appetite: rivers of uninterrupted betterment.

At the heart of this new regime of constant flux is ever tinier specks of computation. We are currently entering the third phase of computing, the Flows.

The initial age of computing borrowed from the industrial age. As Marshall McLuhan observed, the first version of a new medium imitates the medium it replaces. The first commercial computers employed the metaphor of the office. Our screens had a "desktop" and "folders" and "files." They were hierarchically ordered, like much of the industrial age that the computer was overthrowing.

The second digital age overturned the office metaphor and brought us the organizing principle of the web. The basic unit was no longer files but "pages." Pages were not organized into folders, but were arranged into a networked web. The web was a billion hyperlinked pages which contained everything, both stored information and active knowledge. The desktop interface was replaced by a "browser," a uniform window that looked into any and all pages. This web of links was flat.

Now we are transitioning into the third age of computation. Pages and browsers are far less important. Today the prime units are flows and streams. We constantly monitor Twitter streams and the flows of posts on our Facebook wall. We stream photos, movies, and music. News banners stream across the bottom of TVs. We subscribe to YouTube streams, called channels. And RSS feeds from blogs. We are bathed in streams of notifications and updates. Our apps improve in a flow of upgrades. Tags have replaced links. We tag and "like" and "favorite" moments in the streams. Some streams, like Snapchat, WeChat, and WhatsApp, operate totally in the present, with no past or future. They just flow past. If you see something, fine. Then it is gone.

Flowing time has shifted as well. In the first era, tasks were accomplished in batch mode. You got your bills every month. Taxes were all paid on the same day of the year. Telephone service came only in units of 30 days. Items piled up and were dealt with in batches. Then, in the second age, along came the web, and very quickly we expected everything the same day. If we withdrew money from our bank, we expected the deduction to show up in our account that same day, not at the end of the month. If we sent an email, we expected a reply later in the day, not two weeks later, like regular mail. Our cycle time jumped from batch mode to daily mode. This was a big deal. The expectation shifted so fast, many institutions were caught off guard. People ran out of patience waiting to be sent a form they needed to fill out; if they couldn't fill it out that day, they moved on.

Now in the third age, we've moved from daily mode to real time. If we message someone, we expect them to reply instantly. If we spend money, we expect the balance in our account to adjust in real time. Why should medical diagnostics take days to return results instead of immediately? If we take a quiz in class, why shouldn't the score be instant? For news, we demand to know what is happening this very second, not an hour ago. Unless it occurs in real time, it does not exist. The corollary—and this is important—is that in order to operate in real time, everything has to flow.

For instance, watching movies on demand means movies must flow. Like most families who subscribe to Netflix, our family became real-time snobs. Unless a movie was available on streaming, we ignored it. Netflix's DVD catalog is about 10 times larger and higher quality than its streaming catalog, but we'd rather watch a lesser show in real time than wait two days for something better on a DVD. Simultaneity trumps quality.

Real-time books, ditto. In predigital days I bought printed books long before I intended to read them. If I spied an enticing book in a bookstore, I bought it. At first, the internet deepened my hefty backlog because I encountered more and more recommendations online. When the Kindle came along, I switched to primarily purchasing only digital books, but I kept the old habit of purchasing ebooks whenever I encountered a great recommendation. It was so easy! Click, got it. Then I had an epiphany that I am sure others have had as well. If I purchase a book ahead of time, it just sits in the same place that a book I have not bought sits (in the cloud) but in the paid bucket instead of the unpaid bucket. Why not just leave it in the unpaid bucket? So now I don't purchase a book until I am ready to read it in the next 30 seconds. This sort of just-in-time purchasing is the natural consequence of real-time streaming.

In the industrial age, companies did their utmost to save themselves time by increasing their efficiency and productivity. That is not enough today. Now organizations need to save their customers and citizens time. They need to do their utmost to interact in real time. Real time is human time. An ATM gets you cash much faster than waiting for a bank teller—and more efficiently too—but what we really want is instant cash at our fingertips, something like the real-time money offered by the streaming companies of Square, PayPal, Alipay, or Apple Pay. So in order to run in real time, our technological infrastructure needed to liquefy. Nouns needed to be verbs. Fixed solid things became services. Data couldn't remain still. Everything had to flow into the stream of now.

The union of a zillion streams of information intermingling, flowing into each other, is what we call the cloud. Software flows from the cloud to you as a stream of upgrades. The cloud is where your stream of texts go before they arrive on your friend's screen. The cloud is where the parade of movies under your account rests until you call for them. The cloud is the reservoir that songs escape from. The cloud is the seat where the intelligence of Siri sits, even as she speaks to you. The cloud is the new organizing metaphor for computers. The foundational units of this third digital regime, then, are flows, tags, and clouds.

———

The first industry to be steamrolled by the switch to real time and the cloud of copies was music. Perhaps because music itself is so flowing—a stream of notes whose beauty lasts only as long as the stream continues—it was the first to undergo liquidity. As the music industry reluctantly transformed, it revealed a pattern of change that would repeat itself again and again in other media, of books, movies, games, news. Later, the same transformation from fixities to flows began to overturn shopping, transportation, and education. This inevitable shift toward fluidity is now transforming almost every other aspect of society. The saga of music's upgrade to the realm of fluidity will reveal where we are headed.

Music has been altered by technology for more than a century. Early gramophone equipment could make recordings that contained no more than four and a half minutes, so musicians abbreviated meandering works to fit to the phonograph, and today the standard duration of a pop song is four and a half minutes. Cheap industrial reproduction of gramophone recordings 50 years ago unleashed mind-boggling quantities of inexpensive exact copies—and a sense that music was something one consumed.

The grand upset that music is now experiencing—the transformation that pioneers such as Napster and BitTorrent signaled a decade ago—is the shift from analog copies to digital copies. The industrial age was driven by analog copies—exact and cheap. The information age is driven by digital copies—exact and free.

Free is hard to ignore. It propels duplication at a scale that would previously have been unbelievable. The top ten music videos have been watched (for free) over 10 billion times. Of course, it's not just music that is being copied freely. It is text, pictures, video, games, entire websites, enterprise software, 3-D printer files. In this new online world, anything that can be copied will be copied for free.

A universal law of economics says the moment something becomes free and ubiquitous, its position in the economic equation suddenly inverts. When nighttime electrical lighting was new and scarce, it was the poor who burned common candles. Later, when electricity became easily accessible and practically free, our preference flipped and candles at dinner became a sign of luxury. In the industrial age, exact copies became more valuable than a handmade original. No one wants the inventor's clunky "original" prototype refrigerator. Most people want a perfect working clone. The more common the clone, the more desirable it is, since it comes with a network of service and repair outlets.

Now the axis of value has flipped again. Rivers of free copies have undermined the established order. In this new supersaturated digital universe of infinite free digital duplication, copies are so ubiquitous, so cheap—free, in fact—that the only things truly valuable are those that cannot be copied. The technology is telling us that copies don't count anymore. To put it simply: When copies are superabundant, they become worthless. Instead, stuff that can't be copied becomes scarce and valuable.

When copies are free, you need to sell things that cannot be copied. Well, what can't be copied?

Trust, for instance. Trust cannot be reproduced in bulk. You can't purchase trust wholesale. You can't download trust and store it in a database or warehouse it. You can't simply duplicate someone's else's trust. Trust must be earned, over time. It cannot be faked. Or counterfeited (at least for long). Since we prefer to deal with someone we can trust, we will often pay a premium for that privilege. We call that branding. Brand companies can command higher prices for similar products and services from companies without brands because they are trusted for what they promise. So trust is an intangible that has increasing value in a copy-saturated world.

There are a number of other qualities similar to trust that are difficult to copy and thus become valuable in this cloud economy. The best way to see them is to start with a simple question: Why would anyone ever pay for something they could get for free? And when they pay for something they could get for free, what are they purchasing?

In a real sense, these uncopyable values are things that are "better than free." Free is good, but these are better since you'll pay for them. I call these qualities "generatives." A generative value is a quality or attribute that must be generated at the time of the transaction. A generative thing cannot be copied, cloned, stored, and warehoused. A generative cannot be faked or replicated. It is generated uniquely, for that particular exchange, in real time. Generative qualities add value to free copies and therefore are something that can be sold.

Here are eight generatives that are "better than free."

IMMEDIACY

Sooner or later you can find a free copy of whatever you want, but getting a copy delivered to your inbox the moment it is released—or even better, produced—by its creators is a generative asset. Many people go to movie theaters to see films on the opening night, where they will pay a hefty price to see a film that later will be available for free, or almost free, via rental or download. In a very real sense, they are not paying for the movie (which is otherwise "free"); they are paying for the immediacy. Hardcover books command a premium for their immediacy, disguised as a harder cover. First-in-line often commands an extra price for the same good. As a sellable quality, immediacy has many levels, including access to beta versions. Beta versions of apps or software were once devalued because they are incomplete, but we've come to understand that beta versions also possess immediacy, which is valuable. Immediacy is a relative term (minutes to months), but it can be found in every product and service.

PERSONALIZATION

A generic version of a concert recording may be free, but if you want a copy that has been tweaked to sound acoustically perfect in your particular living room—as if it were being performed in your room—you may be willing to pay a lot. You are then not paying for the copy of the concert; you are paying for the generative personalization. The free copy of a book can be custom edited by the publishers to reflect your own previous reading background. A free movie you buy may be cut to reflect the rating you desire for family viewing (no sex, kid safe). In both of these examples, you get the copy free and pay for personalization. Aspirin is basically free today, but an aspirin-based drug tailored to your DNA could be very valuable, and expensive. Personalization requires an ongoing conversation between the creator and consumer, artist and fan, producer and user. It is deeply generative because it is iterative and time-consuming. Marketers call that "stickiness" because it means both sides of the relationship are stuck (invested) in this generative asset and will be reluctant to switch and start over. You can't cut and paste this kind of depth.

INTERPRETATION

As the old joke goes: "Software, free. User manual, $10,000." But it's no joke. A couple of high-profile companies, like Red Hat, Apache, and others make their living selling instruction and paid support for free software. The copy of code, being mere bits, is free. The lines of free code become valuable to you only through support and guidance. A lot of medical and genetic information will go this route in the coming decades. Right now getting a full copy of all your DNA is very expensive ($10,000), but soon it won't be. The price is dropping so fast, it will be $100 soon, and then the next year insurance companies will offer to sequence you for free. When a copy of your sequence costs nothing, the interpretation of what it means, what you can do about it, and how to use it—the manual for your genes, so to speak—will be expensive. This generative can be applied to many other complex services, such as travel and health care.

AUTHENTICITY

You might be able to grab a popular software application for free on the dark net, but even if you don't need a manual, you might want to be sure it comes without bugs, malware, or spam. In that case you'll be happy to pay for an authentic copy. You get the same "free" software, but with an intangible peace of mind. You are not paying for the copy; you are paying for the authenticity. There are nearly an infinite number of variations of Grateful Dead jams around; buying an authentic version from the band itself will ensure you get the one you wanted. Or that it was indeed actually performed by the Dead. Artists have dealt with this problem for a long time. Graphic reproductions such as photographs and lithographs often come with the artist's stamp of authenticity—a signature—to raise the price of the copy. Digital watermarks and other signature technology will not work as copy protection schemes (copies are superconducting liquids, remember?), but they can serve up the generative quality of authenticity for those who care.

ACCESSIBILITY

Ownership often sucks. You have to keep your things tidy, up-to-date, and, in the case of digital material, backed up. And in this mobile world, you have to carry it along with you. Many people, myself included, will be happy to have others tend our "possessions" while we lazily subscribe to them on the cloud. I may own a book or have previously paid for music I treasure, but I'll pay Acme Digital Warehouse to serve me what I want when and how I want it. Most of this material will be available free elsewhere, but it is just not as convenient. With a paid service I have access to free material anywhere, channeled to any of my many devices, with a super user interface. In part, this is what you get with iTunes on the cloud. You pay for conveniently accessible music you could download for free somewhere else. You are not paying for the material; you are paying for the convenience of easy accessibility, without the obligations of maintaining it.

EMBODIMENT

At its core the digital copy is without a body. I am happy to read a digital PDF of a book, but sometimes it is luxurious to have the same words printed on bright white cottony paper bound in leather. Feels so good. Gamers enjoy fighting with their friends online but often crave playing with them in the same room. People pay thousands of dollars per ticket to attend an event in person that is also streamed live on the net. There is no end of ways to counter the intangible world with greater embodiment. There will always be insanely great new display technology that consumers won't have in their home, so they need to move their bodies somewhere else, like to a theater or auditorium. A theater is more likely to be the first to offer laser projection, holographic display, the holodeck itself. And nothing gets embodied as much as music in a live performance, with real bodies. In this accounting, the music is free, the bodily performance expensive. Indeed, many bands today earn their living through concerts, not music sales. This formula is quickly becoming a common one for not only musicians, but even authors. The book is free; the bodily talk is expensive. Live concert tours, live TED talks, live radio shows, pop-up food tours all speak to the power and value of a paid ephemeral embodiment of something you could download for free.

PATRONAGE

Deep down, avid audiences and fans *want* to pay creators. Fans love to reward artists, musicians, authors, actors, and other creators with the tokens of their appreciation, because it allows them to connect with people they admire. But they will pay only under four conditions that are not often met: 1) It must be extremely easy to do; 2) The amount must be reasonable; 3) There's clear benefit to them for paying; and 4) It's clear the money will directly benefit the creators. Every now and then a band or artist will experiment in letting fans pay them whatever they wish for a free copy. This scheme basically works. It's an excellent illustration of the power of patronage. The elusive connection that flows between appreciative fans and the artist is definitely worth something. One of the first bands to offer the option of pay-what-you-want was Radiohead. They discovered they made about $2.26 per download of their 2007 *In Rainbows* album, earning the band more money than all previous albums released on labels combined and spurring several million sales of CDs. There are many other examples of the audience paying simply because they gain an intangible pleasure from it.

DISCOVERABILITY

The previous generatives resided within creative works. Discoverability, however, is an asset that applies to an aggregate of many works. No matter what its price, a work has no value unless it is seen. Unfound masterpieces are worthless. When there are millions of books, millions of songs, millions of films, millions of applications, millions of everything requesting our attention—and most of it free—being found is valuable. And given the exploding numbers of works created each day, being found is increasingly unlikely. Fans use many ways to discover worthy works out of the zillions produced. They use critics, reviewers, brands (of publishers, labels, and studios), and increasingly they rely on other fans and friends to recommend the good stuff. Increasingly they are willing to pay for guidance. Not too long ago *TV Guide* had a million subscribers who paid the magazine to point them to the best shows on TV. These shows, it is worth noting, were free to the viewers. *TV Guide* allegedly made more money than all three major TV networks it "guided" combined. Amazon's greatest asset is not its Prime delivery service but the millions of reader reviews it has accumulated over decades. Readers will pay for Amazon's all-you-can-read ebook service, Kindle Unlimited, even though they will be able to find ebooks for free elsewhere, because Amazon's reviews will guide them to books they want to read. Ditto for Netflix. Movie fans will pay Netflix because their recommendation engine finds gems they would not otherwise discover. They may be free somewhere else, but they are essentially lost and buried. In these examples, you are not paying for the copies, you are paying for the findability.

———

These eight qualities require a new skill set for creators. Success no longer derives from mastering distribution. Distribution is nearly automatic; it's all streams. The Great Copy Machine in the Sky takes care of that. The technical skills of copy protection are no longer useful because you can't stop copying. Trying to prohibit copying, either by legal threats or technical tricks, just doesn't work. Nor do the skills of hoarding and scarcity. Rather, these eight new generatives demand nurturing qualities that can't be replicated with a click of the mouse. Success in this new realm requires mastering the new liquidity.

———

Once something, like music, is digitized, it becomes a liquid that can be flexed and linked. At first glance, when music was initially digitized, it seemed to music executives that audiences were drawn online because of their greed for the free. But in fact, free was only a part of the attraction. And maybe the least important part. Millions of people might have initially downloaded music because it was free, but they then suddenly discovered something even better. Free music was unencumbered. It could merrily migrate to new media, new roles, new corners of the listeners' lives. Thereafter, the sustained rush to download online music came from digitized sound's ever expanding power of flowing.

Before liquidity, music was staid. Our choice as music fans 30 years ago was limited. You could listen to the set sequence of songs the DJs chose to play on a handful of radio stations or you could buy an album and listen to the music in the order the songs were laid on the disk. Or you could purchase a musical instrument and hunt for a favorite piece's sheet music in obscure shops. That was about it.

Liquidity offered new powers. Forget the tyranny of the radio DJ. With liquid music you had the power to reorder the sequence of tunes on an album or among albums. You could shorten a song or draw it out so that it took twice as long to play. You could extract a sample of notes from someone else's song to use yourself. Or you could substitute lyrics in the audio. You could reengineer a piece so that it sounded better on a car woofer. You could—as someone later did—take two thousand versions of the same song and create a chorus from it. The superconductivity of digitalization had unshackled music from its narrow confines on a vinyl disk and thin oxide tape. Now you could unbundle a song from its four-minute package, filter it, bend it, archive it, rearrange it, remix it, mess with it. It wasn't only that it was monetarily free; it was freed from constraints. Now there were a thousand new ways to conjure with those notes.

What counts are not the number of copies but the number of ways a copy can be linked, manipulated, annotated, tagged, highlighted, bookmarked, translated, and enlivened by other media. Value has shifted away from a copy toward the many ways to recall, annotate, personalize, edit, authenticate, display, mark, transfer, and engage a work. What counts is how well the work flows.

At least 30 music streaming services, far more refined than the original Napster, now provide listeners a spectrum of ways to play with the unconfined elements of music. My favorite of these is Spotify because it encapsulates many of the possibilities that a fluid service can provide. Spotify is a cloud containing 30 million tracks of music. I can search that ocean of music to locate the most specific, weirdest, most esoteric song possible. While it plays I click a button and find the song's lyrics displayed. It will make a virtual personal radio station for me from a small selection of my favorite music. I can tweak the station's playlist by skipping songs or downvoting ones I don't want to hear again. This degree of interacting with music would have astounded fans a generation ago. What I'd really like to listen to is the cool music my friend Chris listens to, because he's much more serious about his music discovery than I am. I'd like to share his playlist, which I can subscribe to—meaning that I am actually listening to the music on his playlist, or even to the songs that Chris is listening to right now, in real time. If I really enjoy a particular song I hear on his list—say, an old Bob Dylan basement tape I never heard before—I can copy it onto my own playlist, which I can then share with my friends.

Naturally, this streaming service is free. If I don't want to see or hear the visual and audio ads Spotify displays to pay the artists, I can pay a monthly premium. In the paid version, I can download the digital files to my computer and I can start to remix tracks if I want to. Since it is the age of flowing, I can reach my playlists and personal radio stations from any device, including my phone, or direct the stream into my living room or kitchen speakers. A bunch of other streaming services, such as Sound-Cloud, operate more like an audio YouTube, encouraging its 250 million fans to upload their own music en masse.

Compare this splendid liquidity of options with the few fixed choices available to me just decades ago. No wonder the fans stampeded to the "free" despite the music industry's threat to arrest them.

Where might this go? In the U.S. at this time, 27 percent of music sales are from the streaming mode, and this mode is equal to the sales of CDs. Spotify pays 70 percent of its subscriber revenue to the artists' labels. Despite this initial success, Spotify's music catalog could be bigger because there are still major holdouts, artists like Taylor Swift, who are fighting against streaming. But as the head of the largest music label in the world admitted, the streaming takeover "is inevitable." With flowing streams, music goes from being a noun to a verb once again.

Liquidity brings a new ease in creation. Fungible forms of music encourage amateurs to create their own song and upload it. To invent new formats. New tools, available for free, distributed online, allow music fans to remix tracks, sample sounds, study lyrics, lay down beats with synthetic instruments. Nonprofessionals start making music the same way writers craft a book—by rearranging found elements (words for writers, chords for musicians) into their own point of view.

The superconductivity of digital bits serves as a lubricant to unleash music's untapped options. Music is flowing at digital frequencies into vast new territories. Predigital, music occupied a few niches. Music came on vinyl; it was played on the radio, was heard at concerts, and in a couple hundred films made each year. Postdigital, music is seeping into the rest of our lives, attempting to occupy our entire waking life. Stuffed into the cloud, music rains on us through our earbuds while we exercise, while we are vacationing in Rome, while we wait in line at the DMV. The niches for music have exploded. A renaissance of thousands of documentaries per year demands a soundtrack for each one them. Feature films consume vast quantities of original scores, including thousands of pop songs. Even YouTube creators understand the emotional uplift gained by a soundtrack for their short spots; while most YouTubers recycle prior art without pay, a growing minority see the value in creating custom music. Then there's the hundreds of hours of music required for each big video game. Tens of thousands of commercials need memorable jingles. The latest fashionable media is a podcast, a sort of audible documentary. At least 27 new podcasts launch every day. No decent podcast is without a theme song and, more often, musical scoring for its long-form content. Our entire life is getting a musical soundtrack. All these venues are growth markets, expanding as rapidly as the flows of bits.

Social media were once the domain of texts. The next generation of social media is conducting video and sound. Apps like WeChat, WhatsApp, Vine, Meerkat, Periscope, and many others enable you to share video and audio—in real time—with your network of friends and friends of friends. The tools for quickly making a tune, altering a song, or algorithmically generating music that you share in real time are not far away. Custom music—that is, music that users generate—will become the norm, and indeed it will become the bulk of all music created each year. As music streams, it expands.

As we've learned from the steady democratization of other arts, soon you'll be able to make music without being a musician. One hundred years ago, the only people technically capable of taking a photograph were a few dedicated experimenters. It was an incredibly elaborate and fussy process. It took great technical skill and greater patience before you could coax a picture worth looking at. An expert photographer might take a dozen photos per year. Today anyone with a phone—which is everyone—can instantly take a photo that is a hundred times better in most dimensions than one taken by the average professional a century ago. We are all photographers. Likewise, typography was once an arcane profession. It required many years of expertise to be able to place type on a page in a pleasing and clear way, since there was no WYSIWYG. Maybe a thousand people knew what kerning was. Today they teach kerning in grammar school, and even newbies can accomplish far better typography with digital tools than the average typesetter of old. Same for cartography. The average web hipster can do more with maps today than the best cartographers could manage in the past. So too it will be for music. With new tools accelerating the fluid flow of bits and copies, we will all become musicians.

As music goes, so goes the other media, and then other industries.

Movies repeated the pattern. A movie was once a rare event, one of the most expensive products to produce. It took highly paid guilds of professionals to make even a B-rated movie. Expensive projection equipment was need to view it, so it was troublesome and rare to see a particular one. Then video cameras came along with file sharing networks, and you could watch any film anytime you wanted. Films that you might be able to see once in your life you could now study by watching hundreds of times. A hundred million people became film students, starting to make their own videos and uploading them to YouTube in the billions. Again, the audience pyramid flipped. We are all filmmakers now.

––––––––

The grand move from fixity to flows can be starkly illustrated in the status of books. Books began as authoritative fixed masterpieces. Crafted with great care and reverence, they were machined to last generations. A big fat paper book is the very essence of stability. It sits on a shelf, not moving, not changing, perhaps for thousands of years. Book lover and critic Nick Carr enumerated four ways books embody fixity. Here's my rendition of how books stay:

Fixity of the page—The page stays the same. Whenever you pick it up, it's the same. You can count on it. That means you can reference or cite it, certain it will say the same thing.

Fixity of the edition—No matter which copy of the book you pick up, no matter where or when you purchased it, it will be the same (for that edition), so its text is shared between us. We can discuss a book sure that we are looking at the identical content.

Fixity of the object—With proper care, paper books last a very long time (centuries longer than digital formats), and their text doesn't change as they age.

Fixity of completion—A paper book carries with it a sense of finality and closure. It is done. Complete. Part of the attraction of printed literature is that it is committed to paper, almost like a vow. The author stands upon it.

These four stabilities are very attractive qualities. They make books monumental, something to reckon with. Yet anyone who loves paper books understands that printed volumes are increasingly expensive compared to a digital copy; it's not hard to imagine a time when very few new books will be printed. Today most books are predominantly born as ebooks. Even old books have had their texts scanned and blasted into every corner of the internet, encouraging them to flow freely on the superconducting wires of the net. The four fixities are not present in ebooks, at least not in the versions of ebooks we see today. But while book lovers will miss the fixities, we should be aware that ebooks offer four fluidities to counter them:

Fluidity of the page—The page is a flexible unit. Content will flow to fit any available space, from a tiny screen in a pair of glasses to a wall. It can adapt to your preferred reading device or reading style. The page fits you.

Fluidity of the edition—A book's material can be personalized. Your edition might explain new words if you are a student, or it could skip a recap of the previous books in the series if you've already read them. Customized "my books" are for me.

Fluidity of the container—A book can be kept in the cloud at such low cost that it is "free" to store in an unlimited library and can be delivered instantly anywhere on earth at any time to anyone.

Fluidity of growth—The book's material can be corrected or improved incrementally. The never-done-ness of an ebook (at least in the ideal) resembles an animated creature more than a dead stone, and this living fluidity animates us as creators and readers.

We currently see these two sets of traits—fixity versus fluidity—as opposites, driven by the dominant technology of the era. Paper favors fixity; electrons favor fluidity. But there is nothing to prevent us from inventing a third way—electrons embedded into paper or any other material. Imagine a book of 100 pages, each page a thin flexible digital screen, bound into a spine—that is an ebook too. Almost anything that is solid can be made a little bit fluid, and anything fluid can be embedded into solidness.

What has happened to music, books, and movies is now happening to games, newspapers, and education. The pattern will spread to transportation, agriculture, health care. Fixities such as vehicles, land, and medicines will become flows. Tractors will become fast computers outfitted with treads, land will become a substrate for a network of sensors, and medicines will become molecular information capsules flowing from patient to doctor and back.

These are the Four Stages of Flowing:

1. **Fixed. Rare.** The starting norm is precious products that take much expertise to create. Each is an artisan work, complete and able to stand alone, sold in high-quality reproductions to compensate the creators.

2. **Free. Ubiquitous.** The first disruption is promiscuous copying of the product, duplicated so relentlessly that it becomes a commodity. Cheap, perfect copies are spent freely, dispersed anywhere there is demand. This extravagant dissemination of copies shatters the established economics.

3. **Flowing. Sharing.** The second disruption is an unbundling of the product into parts, each element flowing to find its own new uses and to be remixed into new bundles. The product is now a stream of services issuing from the shared cloud. It becomes a platform for wealth and innovation.

4. **Opening. Becoming.** The third disruption is enabled by the previous two. Streams of powerful services and ready pieces, conveniently

grabbed at little cost, enable amateurs with little expertise to create new products and brand-new categories of products. The status of creation is inverted, so that the audience is now the artist. Output, selection, and quality skyrocket.

These four stages of flowing apply to all media. All genres will exhibit some fluidity. Yet fixity is not over. Most of the good fixed things in our civilization (roads, skyscrapers) are not going anywhere. We will continue to manufacture analog objects (chairs, plates, shoes), but they will acquire a digital essence as well, with embedded chips. (Except for a tiny minority of high-priced handmade artifacts.) The efflorescent blossoming of liquid streams is an additive process, rather than subtractive. The old media forms endure; the new are layered on top of them. The important difference is that fixity is not the *only* option anymore. Good things don't have to be static, unchanging. Or, to put it a different way, the right kind of instability can now be good. The move from stocks to flows, from fixity to fluidity, is not about leaving behind stability. It is about harnessing a wide-open frontier where so many additional options based on mutability are possible. We are exploring all the ways to make things out of ceaseless change and shape-shifting processes.

Here is what a day in the near future looks like. I tap into the cloud to enter the library containing all music, movies, books, VR worlds, and games. I choose music. In addition to songs, I can get parts of the songs as small as a chord. A song's assets are divvied up one channel at a time, which means I can get just the bass or drum track, or just the voices. Or the song with no voices—perfect for karaoke. Tools allow me to stretch or shrink the duration of a song without changing its pitch and melody. Professional tools let me swap instruments in the song I found. One of my favorite musicians releases alternative versions of her songs (for extra cost), and even offers a historical log of every version during its creation.

Movies are similar. The myriad components of each movie are

released in pieces, not just the soundtrack. I can get the sound effects, the special effects (before and after) of each scene, alternative camera views, voice-overs, all in workable shape. Some studios release a whole set of outtakes that can be reedited. Using this wealth of unbundled assets, a subculture of amateur editors reedits released movies in the hopes of bettering the original director. I've done a few here and there in my media classes. Of course, not every director is interested in being reedited, but the demand is so high and the sales of these insider pieces so good that the studios bank on it. Mature-rated movies are reedited for squeaky-clean family versions, or, on the black net, illicit pornographic versions are made from G movies. Many of the hundreds of thousands of documentaries already released are kept updated with material added by viewers, enthusiasts, or the director, as their stories continue.

The streams of video produced and shared by my own mobile devices are born with channels so they can easily be reworked by my friends. Selecting out the background, they insert my buddies into exotic scenes and playfully manipulate the context in a very believable way. Each video posted demands a reply with another video based upon it. The natural response to receiving a clip, a song, a text—either from a friend or from a professional—is not just to consume it, but to act upon it. To add, subtract, reply, alter, bend, merge, translate, elevate to another level. To continue its flow. To maximize the flowing. My media diet may be thought of as streams of pieces, some of which I consume as is, and most of which I engage in to some degree.

———

We have only started flowing. We have begun the four stages of flowing for some types of digital media, but for most we are still at the first stage. So much more of our routines and infrastructure remains to be liquefied, but liquefied and streamed they will be. The steady titanic tilt toward dematerialization and decentralization means that further flows are inevitable. It seems a stretch right now that the most solid and fixed apparatus in our manufactured environment would be transformed into ethereal forces, but the soft will trump the hard. Knowledge will rule atoms. Generative intangibles will rise above the free. Think of the world flowing.

4

SCREENING

I n ancient times culture revolved around the spoken word. The oral skills of memorization, recitation, and rhetoric instilled in oral societies a reverence for the past, the ambiguous, the ornate, and the subjective. We were People of the Word. Then, about 500 years ago, orality was overthrown by technology. Gutenberg's 1450 invention of metallic movable type elevated writing into a central position in the culture. By the means of cheap and perfect copies, printed text became the engine of change and the foundation of stability. From printing came journalism, science, libraries, and law. Printing instilled in society a reverence for precision (of black ink on white paper), an appreciation for linear logic (in a string of sentences), a passion for objectivity (of printed fact), and an allegiance to authority (via authors), whose truth was as fixed and final as a book.

Mass-produced books changed the way people thought. The technology of printing expanded the number of words available, from about 50,000 words in Old English to a million today. More word choices enlarged what could be communicated. More media choices broadened what was written about. Authors did not have to compose scholarly tomes only, but could "waste" inexpensively printed books on heartrending love stories (the romance novel was invented in 1740), or publish memoirs even if they were not kings. People could write tracts to oppose the prevailing consensus, and with cheap printing an unorthodox idea might gain enough influence to topple a king or the pope. In time, the power of authors birthed the reverence for authors, and of authority, and bred a culture of expertise. Perfection was achieved "by the book." Laws were compiled into official tomes, contracts were written down, and nothing was valid unless put into words onto pages. Painting, music, architecture, dance were all important, but the heartbeat of Western culture was the turning pages of a book. By 1910 three quarters of the towns in the United States with more than 2,500 residents had a public library. America's roots spring from documents—the Constitution, the Declaration of Independence, and, indirectly, the Bible. The country's success depended on high levels of literacy, a robust free press, allegiance to the rule of law (found in books), and a common language across a continent. American prosperity and liberty grew out of a culture of reading and writing. We became People of the Book.

But today more than 5 billion digital screens illuminate our lives. Digital display manufacturers will crank out 3.8 billion new additional screens per year. That's nearly one new screen each year for every human on earth. We will start putting watchable screens on any flat surface. Words have migrated from wood pulp to pixels on computers, phones, laptops, game consoles, televisions, billboards, and tablets. Letters are no longer fixed in black ink on paper, but flitter on a glass surface in a rainbow of colors as fast as our eyes can blink. Screens fill our pockets, briefcases, dashboards, living room walls, and the sides of buildings. They sit in front of us when we work—regardless of what we do. We are now People of the Screen.

This has set up the current culture clash between People of the Book and People of the Screen. The People of the Book today are the good hard-working people who make newspapers, magazines, the doctrines of law, the offices of regulation, and the rules of finance. They live by the book, by the authority derived from authors. The foundation of this culture is ultimately housed in texts. They are all on the same page, so to speak.

The immense cultural power of books emanated from the machinery of reproduction. Printing presses duplicated books quickly, cheaply, and faithfully. Even a butcher might own a copy of Euclid's *Elements*, or the Bible, and so printed copies illuminated the minds of citizens beyond the gentry. This same transformative machinery of reproduction was applied to art and music, with equivalent excitation. Printed copies of etchings and woodcuts brought the genius of visual art to the masses. Cheaply copied diagrams and graphs accelerated science. Eventually, inexpensive copies of photography and recorded music spread the reproductive imperative of the book even wider. We could churn out cheap art and music as fast as books.

This reproductive culture has, in the last century or so, produced the greatest flowering of human achievement the world has ever seen, a magnificent golden age of creative works. Cheap physical copies have enabled millions of people to earn a living directly from the sale of their art to the audience, without the weird dynamics of having to rely only on patronage. Not only did authors and artists benefit from this model, but the audience did too. For the first time, billions of ordinary people were able to come in regular contact with a great work. In Beethoven's day, few people ever heard one of his symphonies more than once. With the advent of cheap audio recordings, a barber in Bombay could listen to them all day long.

———

But today most of us have become People of the Screen. People of the Screen tend to ignore the classic logic of books or the reverence for copies; they prefer the dynamic flux of pixels. They gravitate toward movie screens, TV screens, computer screens, iPhone screens, VR goggle screens, tablet screens, and in the near future massive Day-Glo megapixel screens plastered on every surface. Screen culture is a world of constant flux, of endless sound bites, quick cuts, and half-baked ideas. It is a flow of tweets, headlines, instagrams, casual texts, and floating first impressions. Notions don't stand alone but are massively interlinked to everything else; truth is not delivered by authors and authorities but is assembled in real time piece by piece by the audience themselves. People of the Screen make their own content and construct their own truth. Fixed copies don't matter as much as flowing access. Screen culture is fast, like a 30-second movie trailer, and as liquid and open-ended as a Wikipedia page.

On a screen, words move, meld into pictures, change color, and perhaps even change meaning. Sometimes there are no words at all, only pictures or diagrams or glyphs that may be deciphered into multiple meanings. This liquidity is terribly unnerving to any civilization based on text logic. In this new world, fast-moving code—as in updated versions of computer code—is more important than law, which is fixed. Code displayed on a screen is endlessly tweakable by users, while law embossed into books is not. Yet code can shape behavior as much as, if not more than, law. If you want to change how people act online, on the screen, you simply alter the algorithms that govern the place, which in effect polices the collective behavior or nudges people in preferred directions.

People of the Book favor solutions by laws, while People of the Screen favor technology as a solution to all problems. Truth is, we are in transition, and the clash between the cultures of books and screens occurs within us as individuals as well. If you are an educated modern person, you are conflicted by these two modes. This tension is the new norm. It all started with the first screens that invaded our living rooms 50 years ago: the big, fat, warm tubes of television. These glowing altars reduced the time we spent reading to such an extent that in the following decades it seemed as if reading and writing were over. Educators, intellectuals, politicians, and parents in the last half of the last century worried deeply that the TV generation would be unable to write. Screens were blamed for an amazing list of societal ills. But of course we all kept watching. And for a while it did seem as if nobody wrote, or could write, and reading scores trended down for decades. But to everyone's surprise, the cool, interconnected, ultrathin screens on monitors, the new TVs, and tablets at the beginning of the 21st century launched an epidemic of writing that continues to swell. The amount of time people spend reading has almost tripled since 1980. By 2015 more than 60 trillion pages have been added to the World Wide Web, and that total grows by several billion a day. Each of these pages was written by somebody. Right now ordinary citizens compose 80 million blog posts per day. Using their thumbs instead of pens, young people around the world collectively write 500 million quips per day from their phones. More screens continue to expand the volume of reading and writing. The literacy rate in the U.S. has remained unchanged in the last 20 years, but those who can read are reading and writing more. If we count the creation of all words on all screens, you are writing far more per week than your grandmother, no matter where you live.

In addition to reading words on a page, we now read words floating nonlinearly in the lyrics of a music video or scrolling up in the closing credits of a movie. We might read dialog balloons spoken by an avatar in a virtual reality, or click through the labels of objects in a video game, or decipher the words on a diagram online. We should properly call this new activity "screening" rather than reading. Screening includes reading words, but also watching words and reading images. This new activity has new characteristics. Screens are always on; we never stop staring at them, unlike with books. This new platform is very visual and it gradually merges words with moving images. On the screen words zip around and float over images, serving as footnotes or annotations, linking to other words or images. You might think of this new medium as books we watch or television we read.

Despite this resurgence of words, People of the Book reasonably fear that books—and therefore classical reading and writing—will soon die as a cultural norm. If that happens, who will adhere to the linear rationality encouraged by book reading? Who will obey rules if the respect for books of laws is diminished, to be replaced by lines of code that try to control our behavior? Who will pay authors to write when almost everything is available for free on flickering screens? They fear that perhaps only the rich will read books on paper. Perhaps only a few will pay attention to the wisdom on their pages. Perhaps fewer will pay for them. What can replace a book's steadfastness in our culture? Will we simply abandon this vast textual foundation that underlies our current civilization? The old way of reading—not this new way—had an essential hand in creating most of what we cherish about a modern society: literacy, rational thinking, science, fairness, rule of law. Where does that all go with screening? What happens to books?

The fate of books is worth investigating in detail because books are simply the first of many media that screening will transform. First screening will change books, then it will alter libraries of books, then it will modify movies and video, then it will disrupt games and education, and finally screening will change everything else.

———

People of the Book think they know what a book is: It is a sheaf of pages with a spine you can grab. In the past almost anything printed between two covers would count as a book. A list of telephone numbers was called a book, even though it had no logical beginning, middle, or end. A pile of bound blank pages was called a sketchbook; it was unabashedly empty, but it did have two covers and was thus called a book. A gallery of photographs on a stack of pages was a coffee table book even though it contained no words at all.

Today the paper sheets of a book are disappearing. What is left in their place is the conceptual structure of a book—a bunch of symbols united by a theme into an experience that takes a while to complete.

Since the traditional shell of the book is vanishing, it's fair to wonder whether its organization is merely a fossil. Does the intangible container of a book offer any advantages over the many other forms of text available now?

Some scholars of literature claim that a book is really that virtual place your mind goes to when you are reading. It is a conceptual state of imagination that one might call "literature space." According to these scholars, when you are engaged in this reading space, your brain works differently than when you are screening. Neurological studies show that learning to read changes the brain's circuitry. Instead of skipping around distractedly gathering bits, when you read you are transported, focused, immersed.

One can spend hours reading on the web and never encounter this literature space. One gets fragments, threads, glimpses. That is the web's great attraction: miscellaneous pieces loosely joined. But without some kind of containment, these loosely joined pieces spin away, nudging a reader's attention outward, wandering from the central narrative or argument.

A separate reading device seems to help. So far we have tablets, pads, Kindles, and phones. The phone is the most surprising. Commentators had long held that no one would want to read a book on a tiny few-inch-wide glowing screen, but they were wrong. By miles. I and many others happily read books that way. In fact, we don't know yet how small a book-reading screen can go. There is an experimental type of reading called rapid serial visual presentation, which uses a screen only one word wide. As small as a postage stamp. Your eye remains stationary, fixed on one word, which replaces itself with the next word in the text, and then the one after that. So your eye reads a sequence of words "behind" one another rather than in a long string next to one another. A small screen only one word wide can squeeze in almost anywhere, expanding the territory of where we can read.

Over 36 million Kindles and ebook readers with e-ink have been sold. An ebook is a plank that holds a single page. The single page is "turned" by clicking the plank, so that one page dissolves into another page. The reflective e-ink in the later generations of Kindles is as sharp and readable as traditional ink on paper. Yet unlike printed words, with these ebooks you can cut and paste text from the page, follow up hyperlinks, and interact with illustrations.

But there is no reason an ebook has to be a plank. E-ink paper can be manufactured in inexpensive flexible sheets as thin and supple and cheap as paper. A hundred or so sheets can be bound into a sheaf, given a spine, and wrapped between two handsome covers. Now the ebook looks very much like a paper book of old, thick with pages, but it can change its content. One minute the page has a poem on it; the next it has a recipe. Yet you still turn its thin pages (a way to navigate through text that is hard to improve). When you are finished reading the book, you slap the spine. Now the same pages show a different tome. It is no longer a bestselling mystery, but a how-to guide to raising jellyfish. The whole artifact is superbly crafted and satisfying to hold. A well-designed ebook shell may be so sensual it might be worth purchasing a very fine one covered in soft well-worn Moroccan leather, molded to your hand, sporting the most satiny, thinnest sheets. You'll probably have several ebook readers of different sizes and shapes optimized for different content.

Personally, I like large pages in my books. I want an ebook reader that unfolds, origami-like, into a sheet at least as big as a newspaper today. Maybe with as many pages. I don't mind taking a few minutes to fold it back into a pocket-size packet when I am done. I love being able to scan multiple long columns and jump between headlines on one plane. A number of research labs are experimenting with prototypes of books that are projected wide and big via lasers from a pocket device onto a nearby flat surface. A table or a wall becomes the pages of these books, which you turn with hand gestures. The oversize pages provide the old-timey thrill of your eye roaming across multiple columns and many juxtapositions.

The immediate effect of books born digital is that they can flow onto any screen, anytime. A book will appear when summoned. The need to purchase or stockpile a book before you read it is gone. A book is less an artifact and more a stream that flows into your view.

This liquidity is just as true for the creation of books as for consumption. Think of a book in all its stages as a process rather than artifact. Not a noun, but a verb. A book is more "booking" than paper or text. It is a becoming. It is a continuous flow of thinking, writing, researching, editing, rewriting, sharing, socializing, cognifying, unbundling, marketing, more sharing, and screening—a flow that generates a book along the way. Books, especially ebooks, are by-products of the booking process. Displayed on a screen, a book becomes a web of relationships generated by booking words and ideas. It connects readers, authors, characters, ideas, facts, notions, and stories. These relationships are amplified, enhanced, widened, accelerated, leveraged, and redefined by new ways of screening.

Yet the tension between the book and the screen is still being played out. The current custodians of ebooks—screen companies such as Amazon and Google, under orders from the book publishers in New York and with the approval of some bestselling authors—have agreed to cripple the extreme liquidity of ebooks by currently preventing readers from cutting and pasting text easily, or from copying large sections of a book, or from otherwise seriously manipulating the text. Ebooks today lack the fungibility of the ur-text of screening: Wikipedia. But eventually the text of ebooks will be liberated in the near future, and the true nature of books will blossom. We will find out that books never really wanted to be printed telephone directories, or hardware catalogs on paper, or paperback how-to books. These are jobs that screens and bits are much superior at—all that updating and searching—tasks that neither paper nor narratives are suited for. What those kinds of books have always wanted was to be annotated, marked up, underlined, bookmarked, summarized, cross-referenced, hyperlinked, shared, and talked to. Being digital allows them to do all that and more.

We can see the very first glimpses of books' newfound freedom in the Kindles and Fires. As I read a book I can (with some trouble) highlight a passage I would like to remember. I can extract those highlights (with some effort today) and reread my selection of the most important or memorable parts. More important, with my permission, my highlights can be shared with other readers, and I can read the highlights of a particular friend, scholar, or critic. We can even filter the most popular highlights of all readers, and in this manner begin to read a book in a new way. This gives a larger audience access to the precious marginalia of another author's close reading of a book (with their permission), a boon that previously only rare-book collectors witnessed.

Reading becomes social. With screens we can share not just the titles of books we are reading, but our reactions and notes as we read them. Today, we can highlight a passage. Tomorrow, we will be able to link passages. We can add a link from a phrase in the book we are reading to a contrasting phrase in another book we've read, from a word in a passage to an obscure dictionary, from a scene in a book to a similar scene in a movie. (All these tricks will require tools for finding relevant passages.) We might subscribe to the marginalia feed from someone we respect, so we get not only their reading list but their marginalia—highlights, notes, questions, musings.

The kind of intelligent book club discussion as now happens on the book sharing site Goodreads might follow the book itself and become more deeply embedded into the book via hyperlinks. So when a person cites a particular passage, a two-way link connects the comment to the passage and the passage to the comment. Even a minor good work could accumulate a wiki-like set of critical comments tightly bound to the actual text.

Indeed, dense hyperlinking among books would make every book a networked event. The conventional vision of the book's future assumes that books will remain isolated items, independent from one another, just as they are on the shelves in your public library. There, each book is pretty much unaware of the ones next to it. When an author completes a work, it is fixed and finished. Its only movement comes when a reader picks it up to enliven it with his or her imagination. In this conventional vision, the main advantage of the coming digital library is portability—the nifty translation of a book's full text into bits, which permits it to be read on a screen anywhere. But this vision misses the chief revolution birthed by scanning books: In the universal library, no book will be an island. It's all connected.

Turning inked letters into electronic dots that can be read on a screen is simply the first essential step in creating this new library. The real magic will come in the second act, as each word in each book is cross-linked, clustered, cited, extracted, indexed, analyzed, annotated, and woven deeper into the culture than ever before. In the new world of ebooks and etexts, every bit informs another; every page reads all the other pages.

Right now the best we can do in terms of interconnection is to link some text to its source's title in a bibliography or in a footnote. Much better would be a link to a specific passage in another passage in a work, a technical feat not yet possible. But when we can link deeply into documents at the resolution of a sentence, and have those links go two ways, we'll have networked books.

You can get a sense of what this might be like by visiting Wikipedia. Think of Wikipedia as one very large book—a single encyclopedia—which of course it is. Most of its 34 million pages are crammed with words underlined in blue, indicating those words are hyperlinked to concepts elsewhere in the encyclopedia. This tangle of relationships is precisely what gives Wikipedia—and the web—its immense force. Wikipedia is the first networked book. In the goodness of time, each Wikipedia page will become saturated with blue links as every statement is cross-referenced. In the goodness of time, as all books become fully digital, every one of them will accumulate the equivalent of blue underlined passages as each literary reference is networked within that book out to all other books. Each page in a book will discover other pages and other books. Thus books will seep out of their bindings and weave themselves together into one large metabook, the universal library. The resulting collective intelligence of this synaptically connected library allows us to see things we can't see in a single isolated book.

<hr>

The dream of a universal library is an old one: to have in one place all knowledge, past and present. All books, all documents, all conceptual works, in all languages—all connected. It is a familiar hope, in part because long ago we briefly built such a library. The great library at Alexandria, constructed around 300 BC, was designed to hold all the scrolls circulating in the known world. At one time or another, the library held about half a million scrolls, estimated to have been between 30 percent and 70 percent of all books in existence back then. But even before this great library was lost, the moment when all knowledge could be housed in a single building had passed. Since then, the constant expansion of information has overwhelmed our capacity to contain it. For 2,000 years, the universal library, together with other perennial longings like invisibility cloaks, antigravity shoes, and paperless offices, has been a mythical dream that keeps receding further into the infinite future. But might the long-heralded great library of all knowledge really be within our grasp?

Brewster Kahle, an archivist who is backing up the entire internet, says that the universal library is now within reach. "This is our chance to one-up the Greeks!" he chants. "It is really possible with the technology of today, not tomorrow. We can provide all the works of humankind to all the people of the world. It will be an achievement remembered for all time, like putting a man on the moon." And unlike the libraries of old, which were restricted to the elite, this library would be truly democratic, offering every book in every language to every person alive on the planet.

Ideally, in such a complete library we should be able to read any article ever written in any newspaper, magazine, or journal. The universal library should also include a copy of every painting, photograph, film, and piece of music produced by all artists, present and past. Still more, it should include all radio and television broadcasts. Commercials too. Of course, the grand library naturally needs a copy of the billions of dead web pages no longer online and the tens of millions of blog posts now gone—the ephemeral literature of our time. In short, the entire works of humankind, from the beginning of recorded history, in all languages, available to all people, all the time.

This is a very big library. From the days of Sumerian clay tablets until now, humans have "published" at least 310 million books, 1.4 billion articles and essays, 180 million songs, 3.5 trillion images, 330,000 movies, 1 billion hours of videos, TV shows, and short films, and 60 trillion public web pages. All this material is currently contained in all the libraries and archives of the world. When fully digitized, the whole lot could be compressed (at current technological rates) onto 50-petabyte hard disks. Ten years ago you needed a building about the size of a small-town library to house 50 petabytes. Today the universal library would fill your bedroom. With tomorrow's technology, it will all fit onto your phone. When that happens, the library of all libraries will ride in your purse or wallet— if it doesn't plug directly into your brain with thin white cords. Some people alive today are surely hoping that they die before such things happen, and others, mostly the young, want to know what's taking so long.

But the technologies that will bring us a planetary source of all written material will also, in the same gesture, transform the nature of what we now call the book and the libraries that hold them. The universal library and its "books" will be unlike any library or books we have known because, rather than read them, we will screen them. Buoyed by the success of massive interlinking in Wikipedia, many nerds believe that a billion human readers can reliably weave together the pages of old books, one hyperlink at a time. Those with a passion for a special subject, obscure author, or favorite book will, over time, link up its important parts. Multiply that simple generous act by millions of readers, and the universal library can be integrated in full, by fans, for fans.

In addition to a link, which explicitly connects one word or sentence or book to another, readers will also be able to add tags. Smart AI-based search technology overcomes the need for overeducated classification systems so user-generated tags are enough to find things. Indeed, the sleepless smartness in AI will tag text and images automatically in the millions, so that the entire universal library will yield its wisdom to any who seek it.

The link and the tag may be two of the most important inventions of the last 50 years. You are anonymously marking up the web, making it smarter, when you link or tag something. These bits of interest are gathered and analyzed by search engines and AIs in order to strengthen the relationship between the end points of every link and the connections suggested by each tag. This type of intelligence has been indigenous to the web since its birth, but was previously foreign to the world of books. The link and the tag now make screening the universal library possible, and powerful.

We see this effect most clearly in science. Science is on a long-term campaign to bring all knowledge in the world into one vast, interconnected, footnoted, peer-reviewed web of facts. Independent facts, even those that make sense in their own world, are of little value to science. (The pseudo- and parasciences are nothing less, in fact, than small pools of knowledge that are not connected to the large network of science. They are valid only in their own network.) In this way, every new observation or bit of data brought into the web of science enhances the value of all other data points.

Once a book has been integrated into the newly expanded library by means of this linking, its text will no longer be separate from the text in other books. For instance, today a serious nonfiction book will usually have a bibliography and some kind of footnotes. When books are deeply linked, you'll be able to click on the title in any bibliography or any footnote and find the actual book referred to in the footnote. The books referenced in that book's bibliography will themselves be available, and so you can hop through the library in the same way we hop through web links, traveling from footnote to footnote to footnote until you reach the bottom of things.

Next come the words. Just as a web article on, say, coral reefs can have some of its words linked to definitions of fish terms, any and all words in a digitized book can be hyperlinked to other parts of other books. Books, including fiction, will become a web of names and a community of ideas. (You can, of course, suppress links—and their connections—if you don't want to see them, as you might while reading a novel. But novels are a tiny subset of everything that is written.)

Over the next three decades, scholars and fans, aided by computational algorithms, will knit together the books of the world into a single networked literature. A reader will be able to generate a social graph of an idea, or a timeline of a concept, or a networked map of influence for any notion in the library. We'll come to understand that no work, no idea stands alone, but that all good, true, and beautiful things are ecosystems of intertwined parts and related entities, past and present.

Even when the central core of a text is authored by a lone author (as is likely for many fictional books), the auxiliary networked references, discussions, critiques, bibliography, and hyperlinks surrounding a book will probably be a collaboration. Books without this network will feel naked.

At the same time, once digitized, books can be unraveled into single pages or be reduced further, into snippets of a page. These snippets will be remixed into reordered books and virtual bookshelves. Just as the music audience now juggles and reorders songs into new albums or playlists, the universal networked library will encourage the creation of virtual "bookshelves"—a collection of texts, some as short as a paragraph, others as long as entire books—that form a library shelf's worth of specialized information. And as with music playlists, once created, these "bookshelves" or playlists for books will be published and swapped in the public commons. Indeed, some authors will begin to write books to be read as snippets or to be remixed as pages. The ability to purchase, read, and manipulate individual pages or sections is surely what will drive reference books (cookbooks, how-to manuals, travel guides) in the future. You might concoct your own "cookbook shelf" or scrapbook of Cajun recipes compiled from many different sources; it would include web pages, magazine clippings, and entire Cajun cookbooks. This is already starting to happen. The boards of the online site Pinterest allow folks to quickly create scrapbooks of quotes, images, quips, and photos. Amazon currently offers you a chance to publish your own bookshelves ("Listmanias") as annotated lists of books you want to recommend on a particular esoteric subject. And readers are already using Google Books to round up mini libraries on a certain topic—all the books about Swedish saunas, for instance, or the best books on clocks. Once snippets, articles, and pages of books become ubiquitous, shuffleable, and transferable, users will earn prestige and perhaps income for curating an excellent collection.

Libraries (as well as many individuals) aren't eager to relinquish old-fashioned ink-on-paper editions, because the printed book is by far the most durable and reliable long-term storage technology we have. Printed books require no mediating device to read and thus are immune to technological obsolescence. Paper is also extremely stable, compared with, say, hard drives or even CDs. The unchanging edition that anchors an author's original vision without the interference of mashups and remixes will often remain the most valuable edition. In this way, the stability and fixity of a bound book is a blessing. It sits constant, true to its original creation. But it sits alone.

So what happens when all the books in the world become a single liquid fabric of interconnected words and ideas? Four things:

First, works on the margins of popularity will find a small audience larger than the near zero audience they usually have now. It becomes easier to discover that labor-of-love masterpiece on the vegan diets of southern Indian priests. Far out in the long tail of the distribution curve—that extended place of low to no sales where most of the books in the world live—digital interlinking will lift the readership of almost any title, no matter how esoteric.

Second, the universal library will deepen our grasp of history, as every original document in the course of civilization is scanned and cross-linked. That includes all the yellowing newspapers, unused telephone books, dusty county files, and old ledgers now moldering in basements. More of the past will be linked to today, increasing understanding today and appreciation of the past.

Third, the universal networked library of all books will cultivate a new sense of authority. If you can truly incorporate all texts—past and present in all languages—on a particular subject, then you can have a clearer sense of what we as a civilization, a species, do and don't know. The empty white spaces of our collective ignorance are highlighted, while the golden peaks of our knowledge are drawn with completeness. This degree of authority is only rarely achieved in scholarship today, but it will become routine.

Fourth and finally, the full, complete universal library of all works becomes more than just a better searchable library. It becomes a platform for cultural life, in some ways returning book knowledge to the core. Right now, if you mash up Google Maps and monster.com, you get maps of where jobs are located by salary. In the same way, it is easy to see that, in the great networked library, everything that has ever been written about, for example, Trafalgar Square in London could be visible while one stands in Trafalgar Square via a wearable screen like Google Glass. In the same way, every object, event, or location on earth would "know" everything that has ever been written about it in any book, in any language, at any time. From this deep structuring of knowledge comes a new culture of participation. You would be interacting—with your whole body—with the universal book.

Soon a book outside the universal Library of All will be like a web page outside the web, gasping for air. Indeed, the only way for the essence of books to retain their waning authority in our culture is to wire their texts into the universal library. Most new works will be born digital, and they will flow into the universal library as you might add more words to a long story. The great continent of analog books in the public domain, and the 25 million orphan works (neither in print nor in the public domain), will eventually be scanned and connected. In the clash between the conventions of the book and the protocols of the screen, the screen will prevail.

One quirk of networked books is that they are never done, or rather that they become streams of words rather than monuments. Wikipedia is a stream of edits, as anyone who has tried to make a citation to it realizes. A book will be networked in time as well as space.

But why bother calling these things books? A networked book, by definition, has no center and is all edges. Might the unit of the universal library be the sentence, paragraph, or chapter article instead of a book? It might. But there is a power in the long form. A self-contained story, unified narrative, and closed argument has a strong attraction for us. There is a natural resonance that draws a network around it. We'll unbundle books into their constituent bits and pieces and knit those into the web, but the higher-level organization of the book will be the focus for our attention—that remaining scarcity in our economy. A book is an attention unit. A fact is interesting, an idea is important, but only a story, a good argument, a well-crafted narrative is amazing, never to be forgotten. As Muriel Rukeyser said, "The universe is made of stories, not of atoms."

Those stories will play across screens. Everywhere we look, we see screens. The other day I watched clips from a movie as I pumped gas into my car. The other night I saw a movie on the seatback of a plane. Earlier this evening I watched a movie on my phone. We will watch anywhere. Everywhere. Screens playing video pop up in the most unexpected places—like ATM machines and supermarket checkout lines. These ever present screens have created an audience for very short moving pictures, as brief as three minutes, while cheap digital creation tools have empowered a new generation of filmmakers, who are rapidly filling up those screens. We are headed toward screen ubiquity.

The screen demands more than our eyes. The most physically active we get while reading a book is to flip the pages or dog-ear a corner. But screens engage our bodies. Touch screens respond to the ceaseless caress of our fingers. Sensors in game consoles such as the Nintendo Wii track our hands and arms. The controller for a video game screen rewards fast twitching. The newest screens—the ones we view within virtual reality headsets and goggles—elicit whole-body movements. They trigger interaction. Some of the newest screens (such as those on the Samsung Galaxy phone) can follow our eyes to perceive where we gaze. A screen will know what we are paying attention to and for how long. Smart software can now read our emotions as we read the screen and can alter what we see next in response to our emotions. Reading becomes almost athletic. Just as it seemed weird five centuries ago to see someone read silently (literacy was so rare most texts were read aloud for the benefit of all), in the future it will seem weird to watch a screen without some part of our body responding to the content.

Books were good at developing a contemplative mind. Screens encourage more utilitarian thinking. A new idea or unfamiliar fact uncovered while screening will provoke our reflex to do something: to research the term, to query your screen "friends" for their opinions, to find alternative views, to create a bookmark, to interact with or tweet the thing rather than simply contemplate it. Book reading strengthened our analytical skills, encouraging us to pursue an observation all the way down to the footnote. Screening encourages rapid pattern making, associating one idea with another, equipping us to deal with the thousands of new thoughts expressed every day. Screening nurtures thinking in real time. We review a movie while we watch it, or we come up with an obscure fact in the middle of an argument, or we read the owner's manual of a gadget before we purchase it rather than after we get home and discover that it can't do what we need it to do. Screens are instruments of the now.

Screens provoke action instead of persuasion. Propaganda is less effective in a world of screens, because while misinformation travels as fast as electrons, corrections do too. Wikipedia works so well because it removes an error in a single click, making it easier to eliminate a falsehood than to post a falsehood in the first place. In books we find a revealed truth; on the screen we assemble our own myths from pieces. On networked screens everything is linked to everything else. The status of a new creation is determined not by the rating given to it by critics but by the degree to which it is linked to the rest of the world. A person, artifact, or fact does not "exist" until it is linked.

A screen can reveal the inner nature of things. Waving the camera eye of a smartphone over a manufactured product can reveal its price, place of origin, ingredients, and even relevant comments by other owners. With the right app, like Google Translate, a phone's screen can instantly translate the words on a menu or a sign in a foreign country into your home language, in the same font. Or another phone app can augment a stuffed children's toy with additional behaviors and interactions that show up only on the screen. It is as if the screen displays the object's intangible essence.

As portable screens become more powerful, lighter, and larger, they will be used to view more of this inner world. Hold an electronic tablet up as you walk along a street—or wear a pair of magic spectacles or contact lenses—and it will show you an annotated overlay of the real street ahead: where the clean restrooms are, which stores sell your favorite items, where your friends are hanging out. Computer chips are becoming so small, and screens so thin and cheap, that in the next 30 years semitransparent eyeglasses will apply an informational layer to reality. If you pick up an object while peering through these spectacles, the object's (or place's) essential information will appear in overlay text. In this way screens will enable us to "read" everything, not just text.

Yes, these glasses look dorky, as Google Glass proved. It will take a while before their form factor is worked out and they look fashionable and feel comfortable. But last year alone, five quintillion (10 to the power of 18) transistors were embedded into objects other than computers. Very soon most manufactured items, from shoes to cans of soup, will contain a small sliver of dim intelligence, and screens will be the tool we use to interact with this ubiquitous cognification. We will want to watch them.

More important, our screens will also watch us. They will be our mirrors, the wells into which we look to find out about ourselves. Not to see our faces, but our selves. Already millions of people use pocketable screens to input their location, what they eat, how much they weigh, their mood, their sleep patterns, and what they see. A few pioneers have begun lifelogging: recording every single detail, conversation, picture, and activity. A screen both records and displays this database of activities. The result of this constant self-tracking is an impeccable "memory" of their lives and an unexpectedly objective and quantifiable view of themselves, one that no book can provide. The screen becomes part of our identity.

We are screening at all scales and sizes—from the IMAX to the Apple Watch. In the near future we will never be far from a screen of some sort. Screens will be the first place we'll look for answers, for friends, for news, for meaning, for our sense of who we are and who we can be.

————

Someday in the near future my day will be like this:

In the morning I begin my screening while still in bed. I check the screen on my wrist for the time, my wake-up alarm, and also to see what urgent news and weather scrolls by. I screen the tiny panel near the bed that shows messages from my friends. I wipe the messages away with my thumb. I walk to the bathroom. I screen my new artworks—cool photos taken by friends—on the wall; these are more cheerful and sunny than the ones yesterday. I get dressed and screen my outfit in the closet. It shows me that the red socks would look better with my shirt.

In the kitchen I screen the full news. I like the display lying flat, horizontal on the table. I wave my arms over the table to direct the stream of text. I turn to the screens on my cabinets, searching for my favorite cereal; the door screens reveal what is behind them. A screen floating above the refrigerator indicates fresh milk inside. I reach inside and take out the milk. The screen on the side of the milk carton tries to get me to play a game, but I quiet it. I screen the bowl to be sure it is approved clean from the dishwasher. As I eat my cereal, I query the screen on the box to see if it is still fresh and whether the cereal has the genetic markers a friend said it did. I nod toward the table and the news stories advance. When I pay close attention, the screen notices and the news gets more detailed. As I screen deeper, the text generates more links, denser illustrations. I begin screening a very long investigative piece on the local mayor, but I need to take my son to school.

I dash to the car. In the car, my story continues where I left off in the kitchen. My car screens the story for me, reading it aloud as I ride. The buildings we pass along the highway are screens themselves. They usually show advertisements that are aimed at only me, since they recognize my car. These are laser-projected screens, which means they can custom focus images that only I see; other commuters see different images on the same screen. I usually ignore them, except when they show an illustration or diagram from the story I am screening in the car. I screen the traffic to see what route is least jammed this morning. Since the car's navigation learns from other drivers' routes, it mostly chooses the best route, but it is not foolproof yet, so I like to screen where the traffic flows.

At my son's school, I check one of the public wall displays in the side hallway. I raise my palm, say my name, and the screen recognizes me from my face, eyes, fingerprints, and voice. It switches to my personal interface. I can screen my messages if I don't mind the lack of privacy in the hall. I can also use the tiny screen on my wrist. I glance at the messages I want to screen in detail and it expands those. I wave some forward and others I swoosh to the archives. One is urgent. I pinch the air and I am screening a virtual conference. My partner in India is speaking to me. She is screening me in Bangalore. She feels pretty real.

I finally make it to the office. When I touch my chair, my room knows me, and all the screens in the room and on the table are ready for me, picking up from where I left off. The eyes of the screens follow me closely as I conduct my day. The screens watch my hands and eyes a lot. I've become very good in using the new hand-sign commands in addition to typing. After 16 years of watching me work, they can anticipate a lot of what I do. The sequence of symbols on the screens makes no sense to anyone else, just as my colleagues' sequence baffles me. When we are working together, we screen in an entirely different environment. We gaze and grab different tools as we hop and dance around the room. I am a bit old-fashioned and still like to hold smaller screens in my hands. My favorite one is the same leather-cased screen I had in college (the screen is new; just the case is old). It is the same screen I used to create the documentary I did after graduation about the migrants sleeping in the mall. My hands are used to it and it is used to my gestures.

After work I put on augmentation glasses while I jog outside. My running route is clearly in front of me. Overlaid on it I also see all my exercise metrics such as my heart rate and metabolism stats displayed in real time, and I can also screen the latest annotation notes posted virtually on the places I pass. I see the virtual notes in my glasses about an alternative detour left by one of my friends when he jogged this same route an hour earlier, and I see some historical notes stuck to a couple of familiar landmarks left by my local history club (I am a member). One day I may try out the bird identification app that pins bird names on the birds in my glasses when I run through the park.

At home during dinner, we don't allow personal screens at our table, though we screen ambient mood colors in the room. After our meal I will screen to relax. I'll put a VR headset on and explore a new alien city created by an amazing world builder I follow. Or I'll jump into a 3-D movie, or join a realie. Like most students, my son screens his homework, especially the tutorials. Although he likes to screen adventure games, we limit it to one hour during the school week. He can screen a realie in about an hour, speed-screening the whole way, while also scanning messages and photos on three other screens at the same time. On the other hand, I try to slow down. Sometimes I'll screen a book on my lap pad while slow, affirming vistas generated from my archives screen on the walls. My spouse likes nothing better than to lie in bed and screen a favorite story on the ceiling till sleep. As I lay down, I set the screen on my wrist for 6 a.m. For eight hours I stop screening.

5

ACCESSING

A reporter for *TechCrunch* recently observed, "Uber, the world's largest taxi company, owns no vehicles. Facebook, the world's most popular media owner, creates no content. Alibaba, the most valuable retailer, has no inventory. And Airbnb, the world's largest accommodation provider, owns no real estate. Something interesting is happening."

Indeed, digital media exhibits a similar absence. Netflix, the world's largest video hub, allows me to watch a movie without owning it. Spotify, the largest music streaming company, lets me listen to whatever music I want without owning any of it. Amazon's Kindle Unlimited enables me to read any book in its 800,000-volume library without owning books, and PlayStation Now lets me play games without purchasing them. Every year I own less of what I use.

Possession is not as important as it once was. Accessing is more important than ever.

Pretend you live inside the world's largest rental store. Why would you own anything? You can borrow whatever you need within arm's reach. Instant borrowing gives you most of the benefits of owning and few of its disadvantages. You have no responsibility to clean, to repair, to store, to sort, to insure, to upgrade, to maintain. What if this rental store were a magical cupboard, a kind of Mary Poppins carpetbag, where an endless selection of gear was crammed into a bottomless container? All you have to do is knock on the outside and summon an item, and abracadabra—there it is.

Advanced technology has enabled this magical rental store. It's the internet/web/phone world. Its virtual cupboards are infinite. In this maximal rental store the most ordinary citizen can get hold of a good or service as fast as if they possessed it. In some cases, getting hold of it may be faster than finding it in your own "basement." The quality of goods is equal to what you can own. Access is so superior to ownership in many ways that it is driving the frontiers of the economy.

Five deep technological trends accelerate this long-term move toward accessing and away from ownership.

Dematerialization

The trend in the past 30 years has been to make better stuff using fewer materials. A classic example is the beer can, whose basic shape, size, and function have been unchanged for 80 years. In 1950 a beer can was made of tin-coated steel and it weighed 73 grams. In 1972 lighter, thinner, cleverly shaped aluminum reduced the weight to 21 grams. Further ingenious folds and curves introduced yet more reductions in the raw materials such that today the can weighs only 13 grams, or one fifth of its original weight. And the new cans don't need a beer can opener. More benefits for just 20 percent of the material. That's called dematerialization.

On average most modern products have undergone dematerialization. Since the 1970s, the weight of the average automobile has fallen by 25 percent. Appliances tend to weigh less per function. Of course, communication technology shows the clearest dematerialization. Huge PC monitors shrunk to thin flat screens (but the width of our TVs expanded!), while clunky phones on the table become pocketable. Sometimes our products gain many new benefits without losing mass, but the general trend is toward products that use fewer atoms. We might not notice this because, while individual items use less material, we use more items as the economy expands and we thus accumulate more stuff in total. However, the total amount of material we use per GDP dollar is going down, which means we use less material for greater value. The ratio of mass needed to generate a unit of GDP has been falling for 150 years, declining even faster in the last two decades. In 1870 it took 4 kilograms of stuff to generate one unit of the U.S.'s GDP. In 1930 it took only one kilogram. Recently the value of GDP per kilogram of inputs rose from $1.64 in 1977 to $3.58 in 2000—a doubling of dematerialization in 23 years.

Digital technology accelerates dematerialization by hastening the migration from products to services. The liquid nature of services means they don't have to be bound to materials. But dematerialization is not just about digital goods. The reason even solid physical goods—like a soda can—can deliver more benefits while inhabiting less material is because their heavy atoms are substituted by weightless bits. The tangible is replaced by intangibles—intangibles like better design, innovative processes, smart chips, and eventually online connectivity—that do the work that more aluminum atoms used to do. Soft things, like intelligence, are thus embedded into hard things, like aluminum, that make hard things behave more like software. Material goods infused with bits increasingly act as if they were intangible services. Nouns morph to verbs. Hardware behaves like software. In Silicon Valley they say it like this: "Software eats everything."

The decreasing mass of steel in an automobile has already given way to lightweight silicon. An automobile today is really a computer on wheels. Smart silicon enhances a car's engine performance, braking, safety—and all the more true for electric cars. This rolling computer is about to be connected and become an internet car. It will sport wireless connection for driverless navigation, for maintenance and safety, and for the latest, greatest HD 3-D video entertainment. The connected car will also become the new office. If you are not driving in your private space, you will either work or play in it. I predict that by 2025 the bandwidth to a high-end driverless car will exceed the bandwidth into your home.

As cars become more digital, they will tend to be swapped and shared and used in the same social way we swap digital media. The more we embed intelligence and smarts into the objects in our households and offices, the more we'll treat these articles as social property. We'll share aspects of them (perhaps what they are made of, where they are, what they see), which means that we'll think of ourselves as sharing them.

When Amazon founder Jeff Bezos first introduced the Kindle ebook reader in 2007, he claimed it was not a product. He said it was a service selling access to reading material. That shift became more visible seven years later when Amazon introduced an all-you-can-read subscription library of almost a million ebooks. Book fans no longer had to purchase individual books, but could buy access to most books currently published with the purchase of one Kindle. (The price of the basic entry Kindle has been dropping steadily and is headed to be almost free soon.) Products encourage ownership, but services discourage ownership because the kind of exclusivity, control, and responsibility that comes with owner-ship privileges are missing from services.

The switch from "ownership that you purchase" to "access that you subscribe to" overturns many conventions. Ownership is casual, fickle. If something better comes along, grab it. A subscription, on the other hand, gushes a never-ending stream of updates, issues, and versions that force a constant interaction between the producer and the consumer. It is not a onetime event; it's an ongoing relationship. To access a service, a cus-tomer is often committing to it in a far stronger way than when he or she purchases an item. You often get locked into a subscription (think of your mobile phone carrier or cable provider) that is difficult to switch out of. The longer you are with the service, the better it gets to know you; and the better it knows you, the harder it is to leave and start over again. It's almost like being married. Naturally, the producer cherishes this kind of loyalty, but the customer gets (or should get) many advantages for continuing as well: uninterrupted quality, continuous improvements, attentive personalization—assuming it's a good service.

Access mode brings consumers closer to the producer, and in fact the consumer often acts as the producer, or what futurist Alvin Toffler called in 1980 the "prosumer." If instead of owning software, you access software, then you can share in its improvement. But it also means you have been recruited. You, the new prosumer, are encouraged to identify bugs and report them (replacing a company's expensive QA department), to seek technical help from other customers in forums (reducing a company's expensive help desk), and to develop your own add-ons and improvements (replacing a company's expensive development team). Access amplifies the interactions we have with all parts of a service.

The first stand-alone product to be "servicized" was software. Today, selling software as service (SaS) instead of product has become the default mode for almost all software. As an example of SaS, Adobe no longer sells its venerable Photoshop and design tools as discrete products with dated versions, 7.0 or whatever. Instead you subscribe to Photoshop, InDesign, Premiere, etc., or the entire suite of services, and its stream of updates. You sign up and your computer will operate the latest best versions as long as you pay the monthly subscription. This new model entails reorientation by customers comfortable owning something forever.

TV, phones, and software as service are just the beginning. In the last few years we've gotten hotels as service (Airbnb), tools as service (TechShop), clothes as service (Stitch Fix, Bombfell), and toys as service (Nerd Block, Sparkbox). Just ahead are several hundred new startups trying to figure how to do food as service (FaS). Each has its own approach to giving you a subscription to food, instead of purchases. For example, in one scheme you might not buy specific food products; instead, you get access to the benefits of food you need or want—say, certain levels and qualities of protein, nutrition, cuisine, flavors.

Other possible new service realms: Furniture as service; Health as service; Shelter as service; Vacation as service; School as service.

Of course, in all these you still pay; the difference is the deeper relationship that services encourage and require between the customer and the provider.

Real-Time On Demand

Access is also a way to deliver new things in close to real time. Unless something runs in real time, it does not count. As convenient as taxis are, they are often not real time enough. You usually wait too long for one, including the ones you call. And the cumbersome payment procedure at the end is a hassle. Oh, and they should be cheaper.

Uber, the on-demand taxi service, has disrupted the transportation business because it shifts the time equation. When you order a ride, you don't need to tell Uber where you are; your phone does that. You don't have to settle payment at the end; your phone does that. Uber uses the phones of the drivers to locate precisely where they are within inches, so Uber can match a driver closest to you. You can track their arrival to the minute. Anyone who wants to earn some money can drive, so there are often more Uber drivers than taxis, especially during peak demand times. And to make it vastly cheaper (in normal use), if you are willing to share a ride, Uber will match two or three riders going to approximately the same place at the same time to split the fare. These UberPool shared-ride fares might be one quarter the cost of a taxi. Relying on Uber (or its competitors, like Lyft) is a no-brainer.

While Uber is well known, the same on-demand "access" model is disrupting dozens of other industries, one after another. In the past few years thousands of entrepreneurs seeking funding have pitched venture capitalists for an "Uber for X," where X is any business where customers still have to wait. Examples of X include: three different Uber for flowers (Florist Now, ProFlowers, BloomThat), three Uber for laundry, two Uber for lawn mowing (Mowdo, Lawnly), an Uber for tech support (Geekatoo), an Uber for doctor house calls, and three Uber for legal marijuana delivery (Eaze, Canary, Meadow), plus a hundred more. The promise to customers is that you don't need a lawn mower or washing machine or to pick up flowers, because someone else will do that for you—on your command, at your convenience, in real time—at a price you can't refuse. The Uber-like companies can promise this because, instead of owning a building full of employees, they own some software. All the work is outsourced and performed by freelancers (prosumers) ready to work. The job for Uber for X is to coordinate this decentralized work and make it happen in real time. Even Amazon has gotten into the business of matching pros with joes who need home services (Amazon Home Services), from cleaning or setting up equipment to access to goat grazing for lawns.

One reason so much money is flowing into the service frontier is that there are so many more ways to be a service than to be a product. The number of different ways to recast transportation as a service is almost unlimited. Uber is merely one variation. There are dozens more already established, and many more possible. The general approach for entrepreneurs is to unbundle the benefits of transportation (or any X) into separate constituent goods and then recombine them in new ways.

Take transportation as an example. How do you get from point A to point B? Today you can do it in one of eight ways with a vehicle:

1. Buy a car, drive yourself (the default today).
2. Hire a company to drive you to your destination (taxi).
3. Rent a company-owned car, drive yourself (Hertz rental).

4. Hire a peer to drive you to your destination (Uber).

5. Rent a car from a peer, drive yourself (RelayRides).

6. Hire a company to drive you with shared passengers along a fixed route (bus).

7. Hire a peer to drive you with shared passengers to your destination (Lyft Line).

8. Hire a peer to drive you with shared passengers going to a fixed destination (BlaBlaCar).

There are variations upon the variations. Hire the service Shuddle to pick up someone else, like a child at school; some call it an Uber for kids. Sidecar is like Uber, except it runs a reverse auction. You set the price you are willing to pay and let drivers bid to pick you up. There are dozens of emerging companies (like SherpaShare) aimed at serving the drivers instead of riders, helping them manage more than one system and optimizing their routes.

These startups try to exploit inefficiencies in novel ways. They take assets that are unused part-time (such as an empty bedroom, a parked car, unused office space) and match them to people eagerly waiting for them right this second. Employing a distributed network of freelance providers, they can approximate near real-time delivery. Now repeat these same experimental business models in other sectors. Delivery: Let a network of freelancers deliver packages to homes (Uber for FedEx). Design: Let a crowd of designers submit designs, just pay the winner (CrowdSpring). Health care: Coordinate sharing insulin pumps. Real estate: Rent your garage as storage space, or an unused cubicle as office space for a startup (WeWork).

Most of these companies won't make it, even though the idea will thrive. Decentralized businesses are very easy to start, with low cost of entry. If these innovative business models are proven to work, established companies are ready to adapt. There is no reason a rental car company like Hertz can't rent freelancers cars, and no reason why taxi companies can't implement aspects of Uber. But the remixing of benefits will continue to flourish and expand.

Our appetite for the instant is insatiable. The cost of real-time engagement requires massive coordination and degrees of collaboration that were unthinkable a few years ago. Now that most people are equipped with a supercomputer in their pocket, entirely new economic forces are being unleashed. If smartly connected, a crowd of amateurs can be as good as the average solo professional. If smartly connected, the benefits of existing products can be unbundled and remixed in unexpected and delightful ways. If smartly connected, products melt into services that can be accessed continuously. If smartly connected, accessing is the default.

Accessing is not very different from renting. In a rent relationship the renter enjoys many of the benefits of ownership, but without the need for an expensive capital purchase or upkeep. Of course, renters are disadvantaged as well because they may not gain all the benefits of traditional ownership, such as rights of modification, long-term access, or gains in value. The invention of renting was not far behind the invention of property, and today you can rent almost anything. How about women's handbags? Top-of-the-line brand-name handbags sell for $500 or more. Since bags are often matched to outfits or seasonal fashions, a selection of fancy bags can get expensive real quick, so a sizable bag rental business has emerged. Rentals start around $50 per week, depending on the bag's demand. As expected, apps and coordination make renting smoother, more effortless. Renting thrives because, for many uses, it is better than owning. Bags can be swapped to match outfits, returned so one does not need to store them. For short-term uses, sharing ownership makes sense. And for many of the things we will use in the upcoming world, short-term use will be the norm. As more items are invented and manufactured—while the total number of hours in a day to enjoy them remains fixed—we spend less and less time per item. In other words, the long-term trend in our modern lives is that *most* goods and services will be short-term use. Therefore most goods and services are candidates for rental and sharing.

The downside to the traditional rental business is the "rival" nature of physical goods. Rival means that there is a zero-sum game; only one rival prevails. If I am renting your boat, no one else can. If I rent a bag to you, I cannot rent the same bag to another. In order to grow a rental business of physical things, the owner has to keep buying more boats or bags. But, of course, intangible goods and services don't work this way. They are "nonrival," which means you can rent the same movie to as many people who want to rent it this hour. Sharing intangibles scales magnificently. This ability to share on a large scale without diminishing the satisfaction of the individual renter is transformative. The total cost of use drops precipitously (shared by millions instead of one). Suddenly, consumer ownership is not so important. Why own when you get the same real-time utility from renting, leasing, licensing, sharing?

For better or worse, our lives are accelerating, and the only speed fast enough is instant. The speed of electrons will be the speed of the future. Deliberate vacations from this speed will remain a choice, but on average communication technology is biased toward moving everything to on demand. And on demand is biased toward access over ownership.

Decentralization

We are at the midpoint in a hundred-year scramble toward greater decentralization. The glue that holds together institutions and processes as they undergo massive decentering is cheap, ubiquitous communication. Without the ability to remain connected as things spread wide into networks, firms would collapse. That's true, but also slightly backward. It's truer to say that the technological means of instant long-distance communications *enabled* this era of decentralization. That is, once we wrapped the globe in endless circles of wires crossing the deserts and beneath the oceans, decentralization was not only possible, but inevitable.

The consequence of moving away from centralized organization to the flatter worlds of networks is that everything—both tangible and intangible—must flow faster to keep the whole going together. Flows are hard to own; possession seems to just slip through your fingers. Access is a more appropriate stance for the fluid relations that govern a decentralized apparatus.

Nearly every aspect of modern civilization has been flattening down except one: money. Minting money is one of the last jobs left for a central government that most political parties agree is legitimate. It takes a central bank to battle the perennial scourges of counterfeit and fraud. Someone has to regulate the amount of money issued, keep track of the serial numbers, ensure that the money is trusted. A robust currency requires accuracy, coordination, security, enforcement—and an institution that takes responsibility for all those. Thus behind every currency stands a watchful central bank.

But what if you could decentralize money as well? What if you created a distributed currency that was secure, accurate, and trustworthy without centralization? Because if money could be decentralized, then *anything* can be decentralized. But even if you could, why would you?

Turns out you can decentralize money, and the technology to do this may be instrumental in decentralizing many other centralized institutions. The story of how the most centralized aspect of modern life is being decentralized holds lessons for many other unrelated industries.

To begin: I can pay you in cash, and that decentralized transaction is anonymous to a central bank. But moving physical cash around is not practical as our economy goes global. PayPal and other peer-to-peer electronic systems are able to bridge the vast geographical spans on a global economy, but each of its peer-to-peer payments must go through a central database to be sure a dollar is not spent twice or is not fraudulent. Mobile phone and internet companies devised very useful payment schemes for impoverished areas based on a phone app, such as M-Pesa. But until recently even the most advanced e-money system needed a central bank to keep the money honest. Six years ago some shady characters who wanted to sell drugs online with the anonymity of cash were looking for a currency without a government hand. And some admirable characters championing human rights were looking for a money system that would work outside of corrupt or repressive governments, or in places of no governance at all. What they together came up with is Bitcoin.

Bitcoin is a fully decentralized, distributed currency that does not need a central bank for its accuracy, enforcement, or regulation. Since it was launched in 2009, the currency has $3 billion in circulation and 100,000 vendors accepting the coins as payment. Bitcoin may be most famous for its anonymity and the black markets it fueled. But forget the anonymity; it's a distraction. The most important innovation in Bitcoin is its "blockchain," the mathematical technology that powers it. The blockchain is a radical invention that can decentralize many other systems beyond money.

When I send you one U.S. dollar via a credit card or PayPal account, a central bank has to verify that transaction; at the very least it must confirm I had a dollar to send you. When I send you one bitcoin, no central intermediary is involved. Our transaction is posted in a public ledger—called a blockchain—that is distributed to all other bitcoin owners in the world. This shared database contains a long "chain" of the transaction history of all existing bitcoins and who owns them. Every transaction is open to inspection by anyone. That completeness is pretty crazy; it's like every person with a dollar having the complete history of all dollar bills as they move around the world. Six times an hour this open distributed database of coins is updated with all the new transactions of bitcoins; a new transaction like ours must be mathematically confirmed by multiple other owners before it is accepted as legitimate. In this way a blockchain creates trust by relying on mutual peer-to-peer accounting. The system itself—which is running on tens of thousands of citizen computers—secures the coin. Proponents like to say that with bitcoin you trust math instead of governments.

A number of startups and venture capitalists are dreaming up ways to use blockchain technology as a general purpose trust mechanism beyond money. For transactions that require a high degree of trust between strangers, such as real estate escrows and mortgage contracts, this validation was previously provided by a professional broker. But instead of paying a traditional title company a lot of money to verify a complex transaction such as a house sale, an online peer-to-peer block-chain system can execute the exchange for much less cost, or maybe for free. Some blockchain enthusiasts propose creating tools that perform a complicated cascade of transactions that depend on verification (like an import/export deal) using only decentralized automated blockchain technology, thereby disrupting many industries that rely on brokers. Whether Bitcoin itself succeeds, its blockchain innovation, which can generate extremely high levels of trust among strangers, will further decentralize institutions and industries.

An important aspect of the blockchain is that it is a public commons. No one really owns it because, well, everyone owns it. As a creation becomes digital, it tends to become shared; as it becomes shared, it also becomes ownerless. When everyone "owns" it, nobody owns it. That is often what we mean by public property or the commons. I use roads that I don't own. I have immediate access to 99 percent of the roads and high-ways of the world (with a few exceptions) because they are a public commons. We are all granted this street access via our payment of local taxes. For almost any purpose I can think of, the roads of the world serve me as if I owned them. Even better than if I owned them, since I am not in charge of maintaining them. The bulk of public infrastructure offers the same "better than owning" benefits.

The decentralized web/internet is now the central public commons. The good of the web serves me as if I owned it, yet I need to do very little to maintain it. I can summon it anytime, with the snap of a finger. I enjoy the full benefits of its amazing work—answering questions like a genius, navigating like a wizard, entertaining like a pro—without the burdens of ownership, simply by accessing it. (I pay its taxes with my subscriptions for internet access.) The more our society decentralizes, the more important accessing becomes.

Platform Synergy

For a long time there were two basic ways to organize human work: a firm and a marketplace. A firm, such as a company, had definite boundaries, was permission based, and enabled people to increase their efficiency via collaboration more than if they worked outside the firm. A marketplace had more permeable borders, required no permission to participate, and used the "invisible hand" to allot resources most efficiently. Recently a third way to organize work has emerged: the platform.

A platform is a foundation created by a firm that lets other firms build products and services upon it. It is neither market nor firm, but something new. A platform, like a department store, offers stuff it did not create. One of the first widely successful platforms was Microsoft's operating system (OS). Anyone with ambition could build and sell a software program that ran on the OS that Microsoft owned. Many did. Some, like the first spreadsheet, Lotus 1-2-3, prospered tremendously and became mini platforms themselves, birthing plug-ins and other third-party derivatives for their product. Levels of highly interdependent products and services form an "ecosystem" that rests upon the platform. "Ecosystem" is a good description because, just as in a forest, the success of one species (product) depends on the success of others. It is the deep ecological interdependence of a platform that discourages ownership and promotes access instead.

Later, a second generation of platforms acquired more of the attributes of markets, so they were a bit of a market and a firm. One of the first of these was iTunes for the iPhone. Apple, the firm, owned the platform, which also became a marketplace for phone apps. Vendors pitched a virtual stall and sold their apps on iTunes. Apple regulated the market, weeding out junky, exploitative, or nonworking applications. It set rules and protocols. It oversaw the financial exchanges. You could say Apple's new product was the marketplace itself. ITunes was an entire ecosystem of apps constructed on the capabilities built into the phone, and it boomed. Since Apple kept adding ingenious new ways to interact with the phone, including new sensors such as a camera, GPS, and an accelerometer, thousands of novel species of innovations deepened the iPhone ecology.

A third generation of platforms further expanded the power of the marketplaces. Unlike traditional two-sided markets—say, a farmers' market that enables buyers and sellers—a platform ecosystem became a multisided market. A good example of this is Facebook. The firm created some rules and protocols that formed a marketplace where independent sellers (college students) produced their own profiles, which were matched up in a marketplace with their friends. The attention of the students was sold to advertisers. Game companies sold to students. Third-party apps sold to advertisers. Third-party apps sold to other third-party apps. And so on in multiple-way matches. This ecosystem of interdependent species keeps expanding, and will keep expanding as long as Facebook can manage its rules and its own growth as a firm.

The wealthiest and most disruptive organizations today are almost all multisided platforms—Apple, Microsoft, Google, and Facebook. All these giants employ third-party vendors to increase the value of their platform. All employ APIs extensively that facilitate and encourage others to play with it. Uber, Alibaba, Airbnb, PayPal, Square, WeChat, Android are the newer wildly successful multiside markets, run by a firm, that enable robust ecosystems of derivative yet interdependent products and services.

Ecosystems are governed by coevolution, which is a type of biological codependence, a mixture of competition and cooperation. In true ecological fashion, supporting vendors who cooperate in one dimension may also compete in others. For instance, Amazon sells both brand-new books from publishers and, via its ecosystem of used-book stores, cheaper used versions. Used-book vendors compete with one another and with the publishers. The platform's job is to make sure it makes money (and adds value!) whether the parts cooperate or compete. Which Amazon does well.

At almost every level of a platform, sharing is the default—even if it is just the rules of competition. Your success hinges on the success of others. Maintaining the idea of ownership within a platform becomes problematic, because it rests on notions of "private property"; but neither "private" nor "property" has great meaning in an ecosystem. As more is shared, less will act like property. It is not a coincidence that less privacy (constant sharing of intimate lives) and more piracy (disregard of intellectual property) are both breeding on platforms.

However, the move from ownership to access has a price. Part of what you own with ownership is the right—and ability—to modify or control the use of your property. That right of modification is sorely missing in many of today's popular digital platforms. Their standard terms of service forbid it. You are legally restricted as to what you can do with the stuff you access versus what you buy. (To be honest, the ability to modify is also squeezed from classic retail purchases as well—think of those silly shrink-wrap warranties.) But the right and ability to modify and control are present in open source platforms and tools such as the Linux OS or the popular Arduino hardware platform, which is part of their great attraction. The ability and right to improve, personalize, or appropriate what is shared will be a key question in the next iteration of platforms.

Dematerialization and decentralization and massive communication all lead to more platforms. Platforms are factories for services; services favor access over ownership.

Clouds

The movies, music, books, and games that you access all live on clouds. A cloud is a colony of millions of computers that are braided together seamlessly to act as a single large computer. The bulk of what you do on the web and phone today is done on cloud computing. Though invisible, clouds run our digital lives.

A cloud is more powerful than a traditional supercomputer because its core is dynamically distributed. That means that its memory and work is spread across many chips in a massively redundant way. Let's say you were streaming a long movie and suddenly an asteroid smashed one tenth of the machines that made up the cloud. You might not notice any interruption in the movie because the movie file did not reside in any particular machine but was distributed in a redundant pattern across many processors in such a way that the cloud can reconfigure itself if any of those units fail. It's almost like organic healing.

The web is hyperlinked documents; the cloud is hyperlinked data. Ultimately the chief reason to put things onto the cloud is to share their data deeply. Woven together, the bits are made much smarter and more powerful than they could possibly be alone. There is no single architecture for clouds, so their traits are still rapidly evolving. But in general they are huge. They are so large that the substrate of one cloud can encompass multiple football field–size warehouses full of computers located in scores of cities thousands of miles apart. Clouds are also elastic, meaning they can be enlarged or shrunk almost in real time by adding or dropping computers to their network. And because of their inherent redundant and distributed nature, clouds are among the most reliable machines in existence. They can provide the famous five nines (99.999 percent) of near perfect service performance.

A central advantage of a cloud is that the bigger it gets, the smaller and thinner our devices can be. The cloud does all the work, while the device we hold is just the window into the cloud's work. When I look into my phone screen and see a live video stream, I am looking into the cloud. When I flick through book pages on my tablet, I am surfing the cloud. When the face of my smartwatch lights up with a message, it is coming from the cloud. When I flip open my cloudbook laptop, everything that I work on is actually somewhere else, in a cloud.

The ambiguity of where my stuff is and whether it is in fact "mine" can be illustrated by the example of a doc on Google. I usually use the Google Drive app to write a marketing letter. "My" letter appears on my laptop or my phone, but its essence lives in Google's cloud, dispersed across many far-flung machines. A key reason I use Google Drive is its ease of collaboration. A dozen or more collaborators can see that letter on their tablet and work on it—edit, add, delete, modify—as if it were "their" letter. Changes made on any of those copies will appear simultaneously—in real time—on all other copies anywhere in the world. It's kind of miraculous, this distributed cloud existence. Each instance of the letter is much more than a mere copy, a term that suggests an inert reproduction. Rather, each person experiences the distributed copy as the original on their device! Each of the dozen copies is as authentic as the one on my laptop. Authenticity is distributed. This collective interaction and distributed being makes the letter feel less mine and more "ours."

Because it lives on the cloud, Google could easily apply cloud-based

AI to our letter in the future. Besides automatically correcting the spelling and critical grammar, Google might also fact-check the statements in the letter with its new truth-checker called Knowledge-Based Trust. It could add hyperlinks to appropriate terms, and add (with my assent) smart additions that improve it significantly so that it further erodes my sense of possession. More and more of our work and play will leave the isolated realm of individual ownership and migrate to the shared world of the cloud in order to take full advantage of AI and other cloud-based powers.

I already google the cloud for answers instead of trying to remember a URL, or even the spelling of a difficult word. If I re-google my own email (stored in a cloud) to find out what I said (which I do) or rely on the cloud for my memory, where does my "I" end and the cloud start? If all the images of my life, and all the snippets of my interests, and all my notes, and all my chitchat with friends, and all my choices, and all my recommendations, and all my thoughts, and all my wishes—if all this is sitting somewhere, but nowhere in particular, it changes how I think of myself. I am larger than before, but thinner too. I am faster, but at times shallower. I think more like a cloud with fewer boundaries, open to change and full of contradiction. I contain multitudes! This whole mix will be further enhanced with the intelligence of machines and AIs. I will be not just Me Plus, but We Plus.

But what happens if it were to go away? A very diffused me would go away. Friends of mine had to ground their teenager for a serious infraction. They confiscated her cell phone. They were horrified when she became physically ill, vomiting. It was almost as if she'd had an amputation. And in one sense she had. If a cloud company restricts or censors our actions, we'll feel pain. Separation from the comfort and new identity afforded by the cloud will be horrendous and unbearable. If McLuhan is right that tools are extensions of our selves—a wheel an extended leg, a camera an extended eye—then the cloud is our extended soul. Or, if you prefer, our extended self. In one sense, it is not an extended self we own, but one we have access to.

Clouds are mostly commercial so far. There is the Oracle Cloud, IBM's SmartCloud, and Amazon's Elastic Compute Cloud. Google and Facebook run huge clouds internally. We keep coming back to clouds because they are more reliable than we are. They are certainly more reliable than other kinds of machines. My very stable Mac freezes or needs to be rebooted once a month. But Google's cloud platform was down only 14 minutes in 2014, a near insignificant outage for the immense amount of traffic served. The cloud is the Backup. Our life's backup.

All business and much of society today run on computers. Clouds offer computation with astounding reliability, fast speed, expandable depth, and no burdens of maintenance for users. Anyone who owns a computer recognizes those burdens: They take up space, need constant expert attention, and go obsolete instantly. Who would want to own their computer? The answer increasingly is no one. No more than you want to own an electric station, rather than buy electricity from the grid. Clouds enable organizations to access the benefits of computers without the hassle of possession. Expandable cloud computing at discount prices has made it a hundred times easier for a young technology company to start up. Instead of building their own complex computing infrastructure, they subscribe to a cloud's infrastructure. In industry terms, this is infrastructure as service. Computers as service instead of computers as product: access instead of ownership. Gaining cheap access to the best infrastructure by operating on the cloud is a chief reason so many young companies have exploded out of Silicon Valley in the last decade. As they grow fast, they access more of what they don't own. Scaling up with success is easy. The cloud companies welcome this growth and dependence, because the more that people use the cloud and share in accessing their services, the smarter and more powerful their service becomes.

There are practical limits to how gigantic one company's cloud can get, so the next step in the rise of clouds over the coming decades will be toward merging the clouds into one intercloud. Just as the internet is the network of networks, the intercloud is the cloud of clouds. Slowly but surely Amazon's cloud and Google's cloud and Facebook's cloud and all the other enterprise clouds are intertwining into one massive cloud that acts as a single cloud—The Cloud—to the average user or company. A counterforce resisting this merger is that an intercloud requires commercial clouds to share their data (a cloud is a network of linked data), and right now data tends to be hoarded like gold. Data hoards are seen as a competitive advantage, and sharing data freely is hampered by laws, so it will be many years (decades?) before companies learn how to share their data creatively, productively, and responsibly.

There is one final step in the inexorable march toward decentralized access. At the same time we are moving to an intercloud we will also move toward one that is fully decentralized and peer to peer. While the enormous clouds of Amazon, Facebook, and Google are distributed, they are not decentralized. The machines are run by enormous companies, not by a funky network of computers run by your funky peers. But there are ways to make clouds that run on decentralized hardware. We know a decentralized cloud can work, because one did during the student protests in Hong Kong in 2014. To escape the obsessive surveillance the Chinese government pours on its citizens' communications, the Hong Kong students devised a way to communicate without sending their messages to a central cell phone tower or through the company servers of Weibo (the Chinese Twitter) or WeChat (their Facebook) or email. Instead they loaded a tiny app onto their phones called FireChat. Two FireChat-enabled phones could speak to each other directly, via wifi radio, without jumping up to a cell tower. More important, either of the two phones could forward a message to a third FireChat-enabled phone. Keep adding FireChat'd phones and you soon have a full network of phones without towers. Messages that are not meant for one phone are relayed to another phone until they reach their intended recipient. This intensely peer-to-peer variety of network (called a mesh) is not efficient, but it works. That cumbersome forwarding is exactly how the internet operates at one level, and why it is so robust. The result of the FireChat mesh was that the students created a radio cloud that no one owned (and was therefore hard to squelch). Relying entirely on a mesh of their own personal devices, they ran a communications system that held back the Chinese government for months. The same architecture could be scaled up to run any kind of cloud.

There are very good nonrevolutionary reasons to have a decentralized communication system like this. In a large-scale emergency when electrical power is out, a peer-to-peer phone mesh might be the only system working. Each individual phone could be recharged by solar, so a communication system could work without the electrical grid. A phone's range is limited, but you could place small cell phone "repeaters" on building rooftops, also potentially recharged by solar. The repeaters just repeat and forward a message for a longer distance than a phone; they are like nanotowers, but they are not owned by a company. A network of rooftop repeaters and millions of phones would create an ownerless network. More than one startup has been founded to offer this type of mesh service.

An ownerless network upsets many of the regulatory and legal frameworks now in place for our communication infrastructure. Clouds don't have a lot of geography. Whose laws will prevail? The laws of your domicile, the laws of your server's domicile, or the laws of international exchange? Who gets your taxes if all the work is being done in the cloud? Who owns the data, you or the cloud? If all your email and voice calls go through the cloud, who is responsible for what it says? In the new intimacy of the cloud, when you have half-baked thoughts, weird daydreams, should they not be treated differently than what you really believe? Do you own your own thoughts, or are you merely accessing them? All these questions apply not only to clouds and meshes but to all decentralized systems.

———

In the coming 30 years the tendency toward the dematerialized, the decentralized, the simultaneous, the platform enabled, and the cloud will continue unabated. As long as the costs of communications and computation drop due to advances in technology, these trends are inevitable. They are the result of networks of communication expanding till they are global and ubiquitous, and as the networks deepen they gradually displace matter with intelligence. This grand shift will be true no matter where in the world (whether the United States, China, or Timbuktu) they take place. The underlying mathematics and physics remain. As we increase dematerialization, decentralization, simultaneity, platforms, and the cloud—as we increase all those at once, access will continue to displace ownership. For most things in daily life, accessing will trump owning.

Yet only in a science fiction world would a person own nothing at all. Most people will own some things while accessing others; the mix will differ by person. Yet the extreme scenario of a person who accesses all without any ownership is worth exploring because it reveals the stark direction technology is headed. Here is how it will work soon.

I live in a complex. Like a lot of my friends, I choose to live in the complex because of the round-the-clock services I can get. The box in my apartment is refreshed four times a day. That means I can leave my refreshables (like clothes) there and have them replenished in a few hours. The complex also has its own Node where hourly packages come in via drones, robo vans, and robo bikes from the local processing center. I tell my device what I need and then it's in my box (at home or at work) within two hours, often sooner. The Node in the lobby also has an awesome 3-D printing fab that can print just about anything in metal, composite, and tissue. There's also a pretty good storage room full of appliances and tools. The other day I wanted a turkey fryer; there was one in my box from the Node's library in a hour. Of course, I don't need to clean it after I'm done; it just goes back into the box. When my friend was visiting, he decided he wanted to cut his own hair. There were hair clippers in the box in 30 minutes. I also subscribe to a camping gear outfit. Camping gear improves so fast each year, and I use it for only a few weeks or weekends, that I much prefer to get the latest, best, pristine gear in my box. Cameras and computers are the same way. They go obsolete so fast, I prefer to subscribe to the latest, greatest ones. Like a lot of my friends, I subscribe to most of my clothes too. It's a good deal. I can wear something different each day of the year if I want, and I just toss the clothes into the box at the end of the day. They are cleaned and redistributed, and often altered a bit to keep people guessing. They even have a great selection of vintage T-shirts that most other companies don't have. The few special smart-shirts I own are chipped-tagged so they come back to me the next day cleaned and pressed.

I subscribe to several food lines. I get fresh produce directly from a farmer nearby, and a line of hot ready-to-eat meals at the door. The Node knows my schedule, my location on my commute, my preferences, so it's really accurate in timing the delivery. When I want to cook myself, I can get any ingredient or special dish I need. My complex has an arrangement so all the ongoing food and cleaning replenishables appear a day before they are needed in the refrig or cupboard. If I was flush with cash, I'd rent a premium flat, but I got a great deal on my place in the complex because they rent it out anytime I am not there. It's fine with me since when I return it's cleaner than I leave it.

I have never owned any music, movies, games, books, art, or realie worlds. I just subscribe to Universal Stuff. The arty pictures on my wall keep changing so I don't take them for granted. I use a special online service that prepares my walls from my collection on Pinterest. My parents subscribe to a museum service that lends them actual historical works of art in rotation, but that is out of my range. These days I am trying out 3-D sculptures that reconfigure themselves each month so you keep noticing them. Even the toys I had as a kid growing up were from Universal. My mom used to say, "You only play with them for a few months—why own them?" So every couple of months they would go into the box and new toys would show up.

Universal is so smart I usually don't have to wait more than 30 seconds for my ride, even during surges. The car just appears because it knows my schedule and can deduce my plans from my texts, calendar, and calls. I'm trying to save money, so sometimes I'll double or triple up with others on the way to work. There is plenty of bandwidth so we can all screen. For exercise, I subscribe to several gyms and a bicycle service. I get an up-to-date bike, tuned and cleaned and ready at my departure point. For long-haul travel I like these new personal hover drones. They are hard to get when you need them right now since they are so new, but so much more convenient than commercial jets. As long as I travel to complexes in other cities that have reciprocal services, I don't need to pack very much since I can get everything—the same things I normally use—from the local Nodes.

My father sometimes asks me if I feel untethered and irresponsible not owning anything. I tell him I feel the opposite: I feel a deep connection to the primeval. I feel like an ancient hunter-gatherer who owns nothing as he wends his way through the complexities of nature, conjuring up a tool just in time for its use and then leaving it behind as he moves on. It is the farmer who needs a barn for his accumulation. The digital native is free to race ahead and explore the unknown. Accessing rather than owning keeps me agile and fresh, ready for whatever is next.

6

SHARING

B ill Gates once derided advocates for free software with the worst
epithet a capitalist can muster. These folks demanding that software
should be free, he said, were a "new modern-day sort of communists," a malevolent force bent on destroying the monopolistic incentive
that helps support the American dream. Gates was wrong on several
points: For one, free and open source software zealots are more likely to
be political libertarians than commie pinkos. Yet there is some truth to
his allegation. The frantic global rush to connect everyone to everyone all
the time is quietly giving rise to a revised technological version of
socialism.

Communal aspects of digital culture run deep and wide. Wikipedia
is just one notable example of an emerging collectivism. Indeed, not just
Wikipedia but wikis of all sorts. Wikis are a set of documents that are
collaboratively produced; their text can easily be created, added, edited,
or altered by anyone, and by everyone. Different wiki engines operate on
different platforms and OSs with various formatting abilities. Ward Cunningham, who invented the first collaborative web page in 1994, tracks

nearly 150 wiki engines today, each powering myriad sites. Widespread adoption of the share-friendly copyright license known as Creative Commons encourages people to legally allow their own images, text, or music to be used and improved by others without the need for additional permission. In other words, sharing and sampling content is the new default.

There were more than one billion instances of Creative Commons permissions in use in 2015. The rise of ubiquitous file sharing sites such as Tor, where one can find a copy of almost anything that can be copied, is another step toward collaboration since it makes it very easy to begin your creation with something already created. Collaborative commenting sites like Digg, StumbleUpon, Reddit, Pinterest, and Tumblr enable hundreds of millions of ordinary folks to find photos, images, news items, and ideas drawn from professional and friends' sources, and then collectively rank them, rate them, share them, forward them, annotate them, and curate them into streams or collections. These sites act as collaborative filters, promoting the best stuff at the moment. Nearly every day another startup proudly heralds a new way to harness community action. These developments suggest a steady move toward a sort of digital "social-ism" uniquely tuned for a networked world.

We're not talking about your grandfather's political socialism. In fact, there is a long list of past movements this new socialism is not. It is not class warfare. It is not anti-American; indeed, digital socialism may be the newest American innovation. While old-school political socialism was an arm of the state, digital socialism is socialism without the state. This new brand of socialism currently operates in the realm of culture and economics, rather than government—for now.

The type of old-school communism with which Gates hoped to tar the creators of shared software, such as Linux or Apache, was born in an era of centralized communications, top-heavy industrial processes, and enforced borders. Those constraints from early last century gave rise to a type of collective ownership that tried to replace the chaos and failures of a free market with well-thought-out scientific five-year plans devised by a politburo of all-powerful experts. This type of government operating system failed, to put it mildly. The top-down socialism of the industrial era could not keep up with the rapid adaptions, constant innovations, and self-generating energy that democratic free markets offered. Socialistic command economies and centralized communistic regimes were left behind. However, unlike those older strains of red-flag socialism, this new digital socialism runs over a borderless internet, via network communications, generating intangible services throughout a tightly integrated global economy. It is designed to heighten individual autonomy and thwart centralization. It is decentralization extreme.

Instead of gathering on collective farms, we gather in collective worlds. Instead of state factories, we have desktop factories connected to virtual co-ops. Instead of sharing picks and shovels, we share scripts and APIs. Instead of faceless politburos, we have faceless meritocracies where the only thing that matters is getting things done. Instead of national production, we have peer production. Instead of free government rations and subsidies, we have a bounty of free commercial goods and services.

I recognize that the word "socialism" is bound to make many readers twitch. It carries tremendous cultural baggage, as do the related terms "communal," "communitarian," and "collective." I use "socialism" because technically it is the best word to indicate a range of technologies that rely on social interactions for their power. We call social media "social" for this same reason: It is a species of social action. Broadly speaking, social action is what websites and net-connected apps generate when they harness input from very large networks of consumers, or participants, or users, or what we once called the audience. Of course, there's rhetorical danger in lumping so many types of organizations under such an inflammatory heading. But there are no unsoiled terms available in this realm of sharing, so we might as well redeem this most direct one:

social, social action, social media, socialism. When masses of people who own the means of production work toward a common goal and share their products in common, when they contribute labor without wages and enjoy the fruits free of charge, it's not unreasonable to call that new socialism.

What they have in common is the verb "to share." In fact, some futurists have called this economic aspect of the new socialism the "sharing economy" because the primary currency in this realm is sharing.

––––––––

In the late 1990s, activist, provocateur, and aging hippy John Perry Barlow began calling this drift, somewhat tongue in cheek, "dot-communism." He defined dot-communism as a "workforce composed entirely of free agents," a decentralized gift or barter economy without money where there is no ownership of property and where technological architecture defines the political space. He was right about the virtual money since the content that Twitter and Facebook distribute is created by unpaid contributors—that is, users like you. And Barlow was right about the lack of ownership, as explained in the previous chapter. We see sharing economy services such as Netflix and Spotify move audiences away from owning anything. But there is one way in which "socialism" is the wrong word for what is happening: It is not an ideology, not an "ism." It demands no rigid creed. Rather, it is a spectrum of attitudes, techniques, and tools that promote collaboration, sharing, aggregation, coordination, ad hocracy, and a host of other newly enabled types of social cooperation. It is a design frontier and a particularly fertile space for innovation.

In his 2008 book *Here Comes Everybody*, media theorist Clay Shirky suggests a useful hierarchy for sorting through these new social arrangements, ranked by the increasing degree of coordination employed. Groups of people start off simply sharing with a minimum of coordination, and then progress to cooperation, then to collaboration, and finally to collectivism. At each step of this socialism, the amount of additional coordination required enlarges. A survey of the online landscape reveals ample evidence of this phenomenon.

1. Sharing

The online public has an incredible willingness to share. The number of personal photos posted on Facebook, Flickr, Instagram, and other sites is an astronomical 1.8 billion per day. It's a safe bet that the overwhelming majority of these digital photos are shared in some fashion. Then there are status updates, map locations, half-thoughts posted online. Add to this the billions of videos served by YouTube each day and the millions of fan-created stories deposited on fanfic sites. The list of sharing organizations is almost endless: Yelp for reviews, Foursquare for locations, Pinterest for scrapbook pieces. Sharing content is now ubiquitous.

Sharing is the mildest form of digital socialism, but this verb serves as the foundation for all the higher levels of communal engagement. It is the elemental ingredient of the entire network world.

2. Cooperation

When individuals work together toward a large-scale goal, it produces results that emerge at the group level. Not only have amateurs shared billions of photos on Flickr and Tumblr, but they have tagged them with categories, labels, and keywords. Others in the community cull the pictures into sets and boards. The popularity of Creative Commons licensing means that in a sense your picture is my picture. Anyone can use an uploaded photo, just as a communard might use the community wheelbarrow. I don't have to shoot yet another photo of the Eiffel Tower, since the community can provide a better one than I can take myself. That means I can make a presentation, a report, a scrapbook, a website much better because I am not working alone.

Thousands of aggregator sites employ a similar social dynamic for threefold benefit. First, social-facing technology aids a site's users directly by letting them individually tag, bookmark, rank, and archive a found item for their own use. Community members can manage and curate their own collections easier. For instance, on Pinterest, plentiful tags and categories ("pins") enable a user to make very quick and specific scrapbooks that are super easy to retrieve and add to. Second, other users will benefit from an individual's tags, pins, and bookmarks. It makes it easier for them to find similar material. The more tags an image gets in Pinterest, or likes in Facebook, or hashtags on Twitter, the more useful it becomes for others. Third, collective action can create an additional value that can come only from the group as a whole. For instance, a pile of tourist snapshots of the Eiffel Tower, each taken from a different angle by a different tourist at a different time, and each one heavily tagged, can be assembled (using software such as Microsoft's Photosynth) into a stunning 3-D holistic rendering of the whole structure that is far more complex and valuable than the individual shots. In a curious way, this proposition exceeds the socialist promise of "from each according to his ability, to each according to his needs" because it betters what you contribute and delivers more than you need.

Community sharing can unleash astonishing power. Sites like Reddit and Twitter, which let users vote up or retweet the most important items (news bits, web links, comments), can steer public conversation as much, and maybe more, than newspapers or TV networks. Dedicated contributors keep contributing in part because of the wider cultural influence these instruments wield. The community's collective influence is far out of proportion to the number of contributors. That is the whole point of social institutions: The sum outperforms the parts. Traditional socialism ramped up this dynamic via the nation-state. Now digital sharing is decoupled from government and operates at an international scale.

3. Collaboration

Organized collaboration can produce results beyond the achievements of ad hoc cooperation. Just look at any of hundreds of open source software projects, such as the Linux operating system, which underpins most web servers and most smartphones. In these endeavors, finely tuned communal tools generate high-quality products from the coordinated work of thousands or tens of thousands of members. In contrast to the previous category of casual cooperation, collaboration on large, complex projects tends to bring the participants only indirect benefits, since each member of the group interacts with only a small part of the end product. An enthusiast may spend months writing code for a subroutine when the program's full utility is several years away. In fact, the work-reward ratio is so out of kilter from a free-market perspective—the workers do immense amounts of high-market-value work without being paid—that these collaborative efforts make no sense within capitalism.

Adding to the economic dissonance, we've become accustomed to enjoying the products of these collaborations free of charge. Half of all web pages in the world today are hosted on more than 35 million servers running free Apache software, which is open source, community created. A free clearinghouse called 3D Warehouse offers several million complex 3-D models of any form you can image (a boot to a bridge), created and freely swapped by very skilled enthusiasts. Nearly 1 million community-designed Arduinos and 6 million Raspberry Pi computers have been built by schools and hobbyists. Their designs are encouraged to be copied freely and used as the basis for new products. Instead of money, the peer producers who create these products and services gain credit, status, reputation, enjoyment, satisfaction, and experience.

Of course, there's nothing particularly new about collaboration per se. But the new tools of online collaboration support a communal style of production that can shun capitalistic investors and keep ownership in the hands of the producers, who are often the consumers as well.

4. Collectivism

Most people in the West, including myself, were indoctrinated with the notion that extending the power of individuals necessarily diminishes the power of the state, and vice versa. In practice, though, most polities socialize some resources and individualize others. Most free-market national economies have socialized education and policing, while even the most extremely socialized societies today allow some private property. The mix varies around the world.

Rather than viewing technological socialism as one side of a zero-sum trade-off between free-market individualism and centralized authority, technological sharing can be seen as a new political operating system that elevates both the individual and the group at once. The largely unarticulated but intuitively understood goal of sharing technology is this: to maximize both the autonomy of the individual and the power of people working together. Thus, digital sharing can be viewed as a third way that renders irrelevant a lot of the old conventional wisdom.

The notion of a third way is echoed by Yochai Benkler, author of *The Wealth of Networks*, who has probably thought more about the politics of networks than anyone else. "I see the emergence of social production and peer production as an alternative to both state-based and market-based closed, proprietary systems," he writes, noting that these activities "can enhance creativity, productivity, and freedom." The new OS is neither the classic communism of centralized planning without private property nor the undiluted selfish chaos of a free market. Instead, it is an emerging design space in which decentralized public coordination can solve problems and create things that neither pure communism nor pure capitalism can.

Hybrid systems that blend market and nonmarket mechanisms are not new. For decades, researchers have studied the decentralized, socialized production methods of northern Italian and Basque industrial co-ops, in which employees are owners who select management and limit profit distribution independent of state control. But only since the arrival of low-cost, instantaneous, ubiquitous online collaboration has it been possible to migrate the core of those ideas into diverse new realms, like coding enterprise software or writing reference books. More important, the technologies of sharing enable collaboration and collectivism to operate at much larger scales than ever before.

The dream is to scale up this third way beyond local experiments. How big can decentralized collaboration go? Black Duck Open Hub, which tracks the open source industry, lists roughly 650,000 people working on more than half a million projects. That total is three times the size of the General Motors workforce. That is an awful lot of people working for free, even if they're not full-time. Imagine if all the employees of GM weren't paid, yet continued to produce automobiles!

So far, the biggest online collaboration efforts are open source projects, and the largest of them, such as Apache, manage several hundred contributors—about the size of a village. One study estimates that 60,000 person-years of work have poured into the release of Fedora Linux 9, so we have proof that self-assembly and the dynamics of sharing can govern a project on the scale of a town.

Of course, the total census of participants in online collective work is far greater. Reddit, the collaborative filtering site, has 170 million unique visitors per month and 10,000 daily active communities. YouTube claims 1 billion monthly users; they are the workforce that produces the videos that now compete with TV. Nearly 25 million registered users have contributed to Wikipedia; 130,000 of them are designated active. More than 300 million active users have posted on Instagram, and more than 700 million groups participate in Facebook Groups each month.

The number of people who belong to collective software farms or work on projects that require communal decisions still fall short of a nation. But the population of people who live in socialized media is gigantic and still increasing. More than 1.4 billion citizens of Facebook freely share their lives in an informational commune. If it were a nation, Facebook would be the largest country on the planet. Yet the entire economy of this largest country runs on labor that isn't paid. A billion people spend a lot of their day creating content for free. They report on events around them, summarize stories, add opinions, create graphics, make up jokes, post cool photos, and craft videos. They are "paid" in the value of the communication and relations that emerge from 1.4 billion connected verifiable individuals. They are paid by being allowed to stay on the commune.

———

One might expect a lot of political posturing from folks who are constructing an alternative to paid labor. But the coders, hackers, and programmers who design sharing tools don't think of themselves as revolutionaries. The most common motivation for working without pay (according to a survey of 2,784 open source developers) was "to learn and develop new skills." One academic put it this way (paraphrasing): "The major reason for working on free stuff is to improve my own damn software." Basically, overt politics is not practical enough. The internet is less a creation dictated by economics than one dictated by sharing gifts.

However, citizens may not be immune to the politics of a rising tide of sharing, cooperation, collaboration, and collectivism. The more we benefit from such collaboration, the more open we become to socialized institutions in government. The coercive, soul-smashing system that controls North Korea is dead (outside of North Korea); the future is a hybrid that takes cues from both Wikipedia and the moderate socialism of, say, Sweden. There will be a severe backlash against this drift from the usual suspects, but increased sharing is inevitable. There is an honest argument over what to call it, but the technologies of sharing have only begun. On my imgainary Sharing Meter Index we are still at 2 out of 10. There is a whole list of subjects that experts once believed we modern humans would not share—our finances, our health challenges, our sex lives, our innermost fears—but it turns out that with the right technology and the right benefits in the right conditions, we'll share everything.

How close to a noncapitalistic, open source, peer-production society can this movement take us? Every time that question has been asked, the answer has been: closer than we thought. Consider Craigslist. Just classified ads, right? Craigslist is far more than that. It amplified the handy community swap board until it reached a regional audience, then enhanced the ads with pictures. It let the customers do all the work of inputting their own ads and, more important, kept the ads in real time with real-time updates, and to top it off it made them free. National classifieds for free! How could debt-laden corporate newspapers compete with that? Operating without state funding or control, connecting citizens directly to citizens, globally, daily, this mostly free marketplace achieved social good at an efficiency (at its peak it had only 30 employees) that would stagger any government or traditional corporation. Sure, peer-to-peer classified undermines the business model of newspapers, but at the same time it makes an indisputable case that the sharing model is a viable alternative to both profit-seeking corporations and tax-supported civic institutions.

Every public health care expert declared confidently that sharing was fine for photos, but no one would share their medical records. But Patients-LikeMe, where patients pool results of treatments to better their own care, proves that collective action can trump both doctors and privacy scares. The increasingly common habit of sharing what you're thinking (Twitter), what you're reading (StumbleUpon), your finances (Motley Fool Caps), your everything (Facebook) is becoming a foundation of our culture. Doing it while collaboratively building encyclopedias, news agencies, video archives, and software in groups that span continents, with people you don't know and whose class is irrelevant—that makes political socialism seem like the logical next step.

A similar thing happened with free markets over the past century. Every day someone asked: What can markets do better? We took a long list of problems that seemed to require rational planning or paternal government and instead applied marketplace logic. For instance, governments traditionally managed communications, particularly scarce radio airways. But auctioning off the communication spectrum in a marketplace radically increased the optimization of bandwidth and accelerated innovation and new businesses. Instead of a government monopoly distributing mail, let market players like DHL, FedEx, and UPS try it as well. In many cases, a modified market solution worked significantly better. Much of the prosperity in recent decades was gained by unleashing market forces on social problems.

Now we're trying the same trick with collaborative social technology: applying digital socialism to a growing list of desires—and occasionally to problems that the free market couldn't solve—to see if it works. So far, the results have been startling. We've had success in using collaborative technology in bringing health care to the poorest, developing free college textbooks, and funding drugs for uncommon diseases. At nearly every turn, the power of sharing, cooperation, collaboration, openness, free pricing, and transparency has proven to be more practical than we capitalists thought possible. Each time we try it, we find that the power of the sharing is bigger than we imagined.

The power of sharing is not just about the nonprofit sector. Three of the largest creators of commercial wealth in the last decade—Google, Facebook, and Twitter—derive their value from unappreciated sharing in unexpected ways.

The earliest version of Google overtook the leading search engines of its time by employing the links made by amateur creators of web pages.

Each time an ordinary person made a hyperlink on the web, Google calculated that link as a vote of confidence for the linked page and used this vote to give a weight to links throughout the web. So a particular page would get ranked higher for reliability in Google's search results if the pages that linked to it were also linked to pages that other reliable pages linked to. This weirdly circular evidence was not created by Google but was instead derived from the public links shared by millions of web pages. Google was the first to extract value from the shared search results that customers clicked on. Each click by an ordinary user represented a vote for the usefulness of that page. So merely by using Google, the fans themselves made Google better and more economically valuable.

Facebook took something that few people thought was valuable—the web of our friends—and encouraged us to share it, while making it easy for us to share notes and gossip with our newly connected circles. This was a minor benefit to individuals—but immensely complex to accomplish in aggregate. No one had anticipated how powerful this unappreciated sharing would be. Facebook's most powerful asset turned out to be the persistent online identity it needed to create for us in order for this sharing scheme to work. While futuristic products such as Second Life's virtual reality made it easy to share an imaginary version of yourself, Facebook made a lot more money by making it easy to share the authentic version of yourself.

Twitter took a similar tack in exploiting the underappreciated power of simply sharing a 140-character "update." It built a surprisingly huge business in enabling people to share quips, and to collect loose acquaintances. Before then, this level of sharing was not considered worthwhile, let along valuable. Twitter proved that what was merely common glitter to an individual could be made into shared gold when collected and processed in the aggregate, and then organized and disseminated back to the individual and sold in analytic clumps to corporations.

―――――

The shift from hierarchy to networks, from centralized heads to decentralized webs, where sharing is the default, has been the major cultural story of the last three decades—and that story is not done yet. The power of bottom up will still take us further. However, *the bottom is not enough*.

To get to the best of what we want, we need some top-down intelligence too. Now that social technology and sharing apps are all the rage, it's worth repeating: The bottom alone is not enough for what we really want. We need a bit of top-down as well. Every predominantly bottom-up organization that lasts for more than a few years does so because it becomes a hybrid of bottom up plus some top down.

I came to that conclusion through personal experience. I was a co-founding editor of *Wired* magazine. Editors perform a top-down function—we select, prune, solicit, shape, and guide the results of writers. We launched *Wired* in 1993, before the web was invented, and so we had a unique privilege to shape journalism as the web emerged. In fact, *Wired* originated one of the first commercial editorial websites. As we experimented with newly possible ways to create and disseminate news on the web, a key unanswered question was: How much influence should editors wield? It was obvious that new online tools made it easier for the audience not only to contribute writing, but also to edit content as well. The recurring insight was simple: What happens if we turn the old model inside out and have the audience/customers in charge? They would be Toffler's prosumers—consumers who were producers. As innovation expert Larry Keeley once observed: "No one is as smart as everyone." Or as Clay Shirky puts it: "Here comes everybody!" Should we simply let the "everyone" in the audience create the online magazine themselves? Should editors step back and just approve what the wisdom of the crowd creates?

Howard Rheingold, a writer and editor who had been living online for a decade before *Wired*, was one of many pundits who argued that it was now possible to forget the editor. Go with the crowd. Rheingold was at the forefront of the then totally radical belief that content could be assembled entirely from the collective action of amateurs and the audience. Rheingold would later write a book called *Smart Mobs*. We hired him to oversee *HotWired*, *Wired*'s online content site. *HotWired*'s original radical idea was to harness the crowd of readers to write the content that other readers would read. But it was even more radical. The shouts from the back of the bus grew loud declaring that *finally* an author no longer needed editors. No one needed to ask permission to publish. Anyone with an internet connection could post their work and gather an audience; it was the end of publishers controlling the gates. This was a revolution! And since it was a revolution, *Wired* published "A Declaration of the Independence of Cyberspace" announcing the end of old media. New media was certainly spawning rapidly. Among them were the link aggregators such as Slashdot, Digg, and later Reddit that enabled users to vote up or down items and to work together as a collaborative consensus filter, making mutual recommendations based on "others like you."

Rheingold believed that *Wired* would get further faster by unleashing people with strong voices, lots of passion, and the willingness to write without any editors to thwart them. Today we'd call those contributors "bloggers." Or tweeters. In this sense Rheingold was right. The entire content that fuels Facebook and Twitter and all the other social media sites is created by users without editors. A billion amateur citizens unleash libraries of text every second. In fact, the average person online today writes more words in a year than many professional writers of the past. This torrent is unedited, unmanaged, completely bottom up. And the attention given to this immense corpus of prosumer content is significant—it was sold to advertisers for $24 billion in 2015.

I was on the other side of this revolt. My counterargument at the time was that the work of most unedited amateurs was simply not that interesting or consistently reliable. When a million people were writing (or blogging or posting) a million times a week, some intelligent guidance to this flood of available text would be worth a lot. The need for some top-down selection would only increase in value as the amount of user-generated content expanded. Over time, the companies that served user-generated content would have to start to layer bits of editing, selection, and curation to their ocean of material in order to maintain quality and attention to it. There had to be something else beside the pure anarchy of the bottom.

This is true for other types of editors as well. Editors are the middle people—or what are called "curators" today—the professionals between a creator and the audience. These middle folk work at publishers, music labels, galleries, or film studios. While their roles would have to change drastically, the demand for the middle would not go away. Intermediates of some type are needed to shape the cloud of creativity that boils up from the crowd.

Yet, in 1994, who knew? In the spirit of a great experiment, we launched *HotWired*, our online magazine, as a primarily user-generated content site. It didn't work. We quickly began adding some editorial oversight and editorially commissioned articles. Users could submit material, but it needed to be edited before publishing. Every decade since then a few commercial news organizations tried this experiment again. *The Guardian* tried to harness readers' reports on a news blog, but it died after two years. *OhMyNews* in South Korea did better than most and ran a reader-written news organization for years before it was returned to editors in 2010. The veteran business magazine *Fast Company* signed up 2,000 blogging readers to report articles sans editors, but closed the experiment after a year and now relies again on readers to suggest ideas for editors to assign. This hybrid of user-generated and editor-enhanced is quite common. Facebook has already started to filter, via intelligent algorithms, the bottom-up flood of news to your feed. It will only continue to add layers of intermediation, as will other bottom-up services.

If one looks hard and honestly, even the supposed paragon of user-generated content—Wikipedia itself—is far from pure bottom-up. In fact, Wikipedia's open-to-anyone process contains an elite in the back room. The more articles someone edits, the more likely their edits will endure and not be undone, which means that over time veteran editors find it easier to make edits that stick, which means that the process favors those few editors who devote lots of time over many years. These persistent old hands act as a type of management, supplying a thin layer of editorial judgment and continuity to this open ad hocracy. In fact, this relatively small group of self-appointed editors is why Wikipedia continues to work and grow into its third decade.

When a community cooperates to write an encyclopedia, as it does in Wikipedia, no one is held responsible if it fails to reach consensus on an article. That gap is simply an imperfection that may or may not get fixed in time. These failures don't endanger the enterprise as a whole. The aim of a collective, on the other hand, is to engineer a system where self-directed peers take responsibility for critical processes and where difficult decisions, such as sorting out priorities, are decided by all participants. Throughout history, countless small-scale collectivist groups have tried this decentralized operating mode in which the executive function is not held at the top. The results have not been encouraging; very few communes have lasted longer than a few years.

Indeed, a close examination of the governing kernel of, say, Wikipedia, Linux, or OpenOffice shows that these efforts are a bit further from the collectivist nirvana than appears from the outside. While millions of writers contribute to Wikipedia, a smaller number of editors (around 1,500) are responsible for the majority of the editing. Ditto for collectives that write code. A vast army of contributions is managed by a much smaller group of coordinators. As Mitch Kapor, founding chair of the Mozilla open source code factory, observed, "Inside every working anarchy, there's an old-boy network."

This isn't necessarily a bad thing. Some types of collectives benefit from a small degree of hierarchy while others are hurt by it. Platforms like the internet, Facebook, or democracy are intended to serve as an arena for producing goods and delivering services. These infrastructural courtyards benefit from being as nonhierarchical as possible, minimizing barriers to entry and distributing rights and responsibilities equally. When powerful actors dominate in these systems, the entire fabric suffers. On the other hand, organizations built to create products rather than platforms often need strong leaders and hierarchies arranged around timescales: Lower-level work focuses on hourly needs; the next level on jobs that need to be done today. Higher levels focus on weekly or monthly chores, and levels above (often in the CEO suite) need to look out ahead at the next five years. The dream of many companies is to graduate from making products to creating a platform. But when they do succeed (like Facebook), they are often not ready for the required transformation in their role; they have to act more like governments than companies in keeping opportunities "flat" and equitable, and hierarchy to a minimum.

In the past, constructing an organization that exploited hierarchy yet maximized collectivism was nearly impossible. The costs of managing so many transactions was too dear. Now digital networking provides the necessary peer-to-peer communication cheap. The net enables a product-focused organization to function collectively by keeping its hierarchy from fully taking over. For instance, the organization behind MySQL, an open source database, is not without some hierarchy, but it is far more collectivist than, say, the giant database corporation Oracle. Likewise, Wikipedia is not exactly a bastion of equality, but it is vastly more collectivist than the *Encyclopaedia Britannica*. The new collectives are hybrid organizations, but leaning far more to the nonhierarchical side than most traditional enterprises.

It's taken a while but we've learned that while top down is needed, not much of it is needed. The brute dumbness of the hive mind is the raw food ingredients that smart design can chew on. Editorship and expertise are like vitamins for the food. You don't need much of them, just a trace even for a large body. Too much will be toxic, or just flushed away. The proper dosage of hierarchy is just barely enough to vitalize a very large collective.

The exhilarating frontier today is the myriad ways in which we can mix large doses of out-of-controlness with small elements of top-down control. Until this era, technology was primarily all control, all top down. Now it can contain both control and messiness. Never before have we been able to make systems with as much messy quasi-control in them. We are rushing into an expanding possibility space of decentralization and sharing that was never accessible before because it was not technically possible. Before the internet there was simply no way to coordinate a million people in real time or to get a hundred thousand workers collaborating on one project for a week. Now we can, so we are quickly exploring all the ways in which we can combine control and the crowd in innumerable permutations.

However, a massively bottom-up effort will take us only partway to our preferred destination. In most aspects of life we want expertise. But we are unlikely to get the level of expertise we want with no experts at all.

That's why it should be no surprise to learn that Wikipedia continues to evolve its process. Each year more structure is layered in. Controversial articles can be "frozen" by top editors so they can no longer be edited by any random person, only designated editors. There are more rules about what is permissible to write, more required formatting, more approval needed. But the quality improves too. I would guess that in 50 years a significant portion of Wikipedia articles will have controlled edits, peer review, verification locks, authentication certificates, and so on. That's all good for us readers. Each of these steps is a small amount of top-down smartness to offset the dumbness of a massively bottom-up system.

Yet if the hive mind is so dumb, why bother with it at all?

Because as dumb as it is, it is smart enough for a lot of work.

In two ways: First, the bottom-up hive mind will always take us much further than we imagine. Wikipedia, though not ideal, is far, far better than anyone believed it could be. It keeps surprising us in this regard. Netflix's personal recommendations derived from what millions of other people watch succeeded beyond what most experts expected. In terms of range of reviews, depth, and reliability, they are more useful than the average human movie critic. EBay's swap meet of virtual strangers was not supposed to work at all, but while not perfect, it is much better than most retailers believed was possible. Uber's peer-to-peer on-demand taxi service works so well it surprised even some of its funders. Given enough time, decentralized connected dumb things can become smarter than we think.

Second, even though a purely decentralized power won't take us all the way, it is almost always the best way to start. It's fast, cheap, and out of control. The barriers to start a new crowd-powered service are low and getting lower. A hive mind scales up wonderfully smoothly. That is why there were 9,000 startups in 2015 trying to exploit the sharing power of decentralized peer-to-peer networks. It does not matter if they morph over time. Perhaps a hundred years from now these shared processes, such as Wikipedia, will be layered up with so much management that they'll resemble the old-school centralized businesses. Even so, the bottom up was still the best way to start.

———

We live in a golden age now. The volume of creative work in the next decade will dwarf the volume of the last 50 years. More artists, authors, and musicians are working than ever before, and they are creating significantly more books, songs, films, documentaries, photographs, artworks, operas, and albums every year. Books have never been cheaper, and more available, than today. Ditto for music, movies, games, and every kind of creative content that can be digitally copied. The volume and variety of creative works available have skyrocketed. More and more of civilization's past works—in all languages—are no longer hidden in rare-book rooms or locked up in archives, but are available a click away no matter where you live. The technologies of recommendation and search have made it super easy to locate the most obscure work. If you want 6,000-year-old Babylonian chants accompanied by the lyre, there they are.

At the same time, digital creation tools have become so ubiquitous that it requires very few resources, or special skills, to produce a book, or a song, or a game, or even a video. Just to prove a point, recently an ad agency shot a very slick TV commercial using smartphones. Legendary painter David Hockney created a popular set of paintings using an iPad. Famous musicians use off-the-shelf hundred-dollar keyboards to record hit songs. More than a dozen unknown authors together have sold millions of self-published ebooks, using nothing more than a dirt-cheap laptop. Speedy global interconnection has produced the largest mass audience yet. On the internet the biggest hits keep getting bigger. The Korean pop dance video "Gangnam Style" has been watched 2.4 billion times and is still going. This size audience has never been seen on the planet before.

While the self-made bestsellers get all the headlines, the real news lies in the other direction. The digital age is the age of non-bestsellers—the underappreciated, the forgotten. Because of sharing technologies, the most obscure interest is no longer obscure; it is one click away. The fast-flowing penetration of the internet into all households, and recently into all pockets via a phone, has put an end to the domination of the mass audience. Most of the time, for most creations, it's a world of niche fulfillment. Left-handed tattoo artists can find each other and share stories and esoteric techniques. People who find whispering sexy (and it turns out many do) can watch whispering videos produced and shared by like-minded whispering folks.

Each of these tiny niches is micro-small, but there are tens of millions of niches. And even though each of those myriad niche interests might attract only a couple of hundred fans, a potential new fan merely has to google to find them. In other words, it becomes as easy to find a particular niche interest as to find a bestseller. Today we are not surprised by a microcommunity sharing an unlikely passion; we are surprised if there is *not* one. We can head out in the wilds of Amazon, Netflix, Spotify, or Google with pretty good confidence that we will uncover someone who has anticipated our most remote interests with a finished work or forum. Each niche is just one step away from a bestselling niche.

Today the audience is king. But what about the creators? Who will pay them in this sharing economy? How will their creative acts be financed if the middle is gone? The surprising answer is: another new sharing technology. No method has been as beneficial to creators as crowdfunding. In crowdfunding the audience funds the work. The fans collectively finance their favorites. The technology of sharing enables the power of one fan who is willing to prepay an artist or author to be aggregated (with little effort) together with hundreds of other fans into a significant pool of money.

The most renowned crowdfunder is Kickstarter, which in the seven years since it was launched has enabled 9 million fans to fund 88,000 projects. Kickstarter is one of about 450 crowdfunding platforms worldwide; others, such as Indiegogo, are almost as prolific. Altogether, crowdfunding platforms raise more than $34 billion each year for projects that would not have been funded in any other way.

In 2013, I was one of about 20,000 people who raised money from fans on Kickstarter. A few friends and I created a full-color graphic novel—or what used to be called a comic book for grown-ups. We calculated we needed $40,000 to pay writers and artists to create and print the second volume of our story, called *The Silver Cord*. So we went onto Kickstarter and made a short video pitch for what we wanted the money for.

Kickstarter runs an ingenious escrow service so that the full grant (in our case $40,000) is not handed over to the creators until and unless the total amount is raised. If the drive is even a dollar short at the end of 30 days, the money is returned immediately to the funders and the fundraisers (us) get nothing. This protects the fans, since an insufficiently funded project is doomed to fail; it also employs the classic network economics of turning your fans into your chief *marketers*, since once they contribute they become motivated to make sure you reach your goal by recruiting their friends to your campaign.

Occasionally, unexpectedly popular fan-financed Kickstarter projects may pile on an additional $1 million above the goal. The highest grossing Kickstarter campaign raised $20 million for a digital watch from its future fans. Approximately 40 percent of all projects succeed in reaching their funding goal.

Each of the 450 or so fan-funding platforms tweak their rules to cater to different groups of creatives or to emphasize different results. Crowdfunding sites can optimize for musicians (PledgeMusic, SellaBand), nonprofits (Fundly, FundRazr), medical emergencies (GoFundMe, Rally), and even science (Petridish, Experiment). A few sites (Patreon, Subbable) are engineered to supply continuous support to an ongoing project like a magazine or video channel. A couple platforms (Flattr, Unglue) use fans to fund work that has already been released.

But by far the most potent future role for crowdsharing is in fan base equity. Rather than invest into a product, supporters invest into a company. The idea is to allow fans of a company to purchase *shares* in the company. This is exactly what you do when you buy shares of stock on the stock market. You are part of a crowdsourced ownership. Each of your shares is some tiny fraction of the whole enterprise, and the collected money raised by public shares is used to grow the business. Ideally, the company is raising money from its own customers, although in reality big pension and hedge funds are the bulk buyers. Heavy regulation and intense government oversight of public companies offer some guarantee to the average stock buyer, making it so anyone with a bank account can buy stock. But risky startups, solo creators, crazy artists, or a duo in their garage would not withstand the kind of paperwork and layers of financial bureaucracy ordinarily applied to public companies. Every year a precious few well-funded companies will attempt an initial public offering (IPO), but only after highly paid lawyers and accountants scour the business in an expensive due diligence scrub. An open peer-to-peer scheme that enabled anyone to offer to the public ownership shares in their company (with some regulation) would revolutionize business. Just as we have seen tens of thousands of new products that would not have existed except by crowdfunding techniques, the new methods of equity sharing would unleash tens of thousands of innovative businesses that could not be born otherwise. The sharing economy would now include ownership sharing.

The advantages are obvious. If you have an idea, you can seek investment from anyone else who sees the same potential as you do. You don't need the permission of bankers, or the rich. If you work hard and succeed, your backers will prosper with you. An artist might use fans' investments to build a company that sold her works over the long term. Or two guys in a garage with an amazing gizmo might be able to leverage that into an ongoing enterprise process that makes more gizmos instead of having to Kickstart each one. The disadvantages are obvious as well. Without some kind of vetting, policing, and enforcement, peer-to-peer investing would be a magnet for huskers and scams. The con artists would offer some kind of glorious returns, take your money, and plead failure. Grannies might lose their life savings. But just as eBay used new innovative technology to solve the old problem of fraud between invisible strangers selling to invisible strangers, the dangers of equity crowdsharing can be minimized with technical innovations such as insurance pools, escrow accounts, and other types of technologically induced trust.

Two early attempts at equity crowdfunding in the U.S., SeedInvest and FundersClub, still rely on rich "qualified investors" and are awaiting a change in U.S. law that would legalize equity crowdfunding for ordinary citizens in early 2016.

Why stop there? Who would have believed that poor farmers could secure $100 loans shared from perfect strangers on the other side of the planet—and pay them back? That is what Kiva does with peer-to-peer lending. Several decades ago international banks discovered they had better repayment rates when they lent small amounts to the poor than when they lent big amounts to rich state governments. It was safer to lend money to the peasants in Bolivia than to the government of Bolivia. This microfinancing of a few hundred dollars applied many tens of thousands of times would also jump-start a developing economy from the bottom. Loan a poor woman $95 to buy supplies to launch a street food cart and the benefits of her stable income would ripple up through her children, the local economy, and quickly build a base for more complex startups. It was the most efficient development strategy invented yet. Kiva took the next step in sharing and turned microfinancing into peer-to-peer lending by enabling anyone, anywhere to make a microfinance loan. So you, sitting at Starbucks, could now lend $120 to a specific individual Bolivian woman who plans to buy wool to start a weaving business. You could follow her progress until she paid you back, at which time you could relend the money to someone else. Since Kiva's launch in 2005, over 2 million people have lent more than $725 million in microfinance loans via its sharing platform. The payback rate is about 99 percent. That is a strong encouragement to lend again.

If that works in developing countries with Kiva, why not install peer-to-peer lending in developed countries? Two web-based companies, Prosper and Lending Club, do that. They match up ordinary middle-class citizen borrowers with ordinary citizen lenders willing to loan their scheme at a decent interest rate. As of 2015, these two largest peer-to-peer lending companies have facilitated more than 200,000 loans worth more than $10 billion.

Innovation itself can be crowdsourced. The Fortune 500 company General Electric was concerned that its own engineers could not keep up with the rapid pace of invention around them, so it launched the platform Quirky. Anyone could submit online an idea for a great new GE product. Once a week, the GE staff voted on the best idea that week and would set to work making it real. If an idea became a product, it would earn money for the idea maker. To date GE has launched over 400 new products from this crowdsourced method. One example is the Egg Minder, an egg holder in your refrigerator that sends you a text when it's time to reorder your eggs.

Another popular version of crowdsourcing appears, at first, to be less about collaboration and more about competition. A commercial need prompts a contest for the best solution. A company offers a payment prize to the best solution selected among a crowd of entrants. For instance, Netflix announced an award of $1 million to the programmers who could invent an algorithm that recommended movies 10 percent better than the algorithm they had. Forty thousand groups submitted very good solutions that improved the performance, but only one team achieved the goal and won the prize. The others had worked for free. Sites such as 99Designs, TopCoder, or Threadless will run a contest for you. Say you need a logo. You offer a fee for the best design. The higher your fee, the more designers will participate. Out of the hundred design sketches submitted, you pick the one you like best and pay its designer. But the open platform means that everyone's work is on view, so each contestant is building upon the creativity of others and trying to outperform them. From the client's point of view, the crowd has generated a design that is probably way better than the one they could have got from just one designer in that price category.

Can a crowd make a car? Yep. Local Motors, based in Phoenix, employs an open source method to design and manufacture low-volume customized performance (fast) cars. A community of 150,000 car fanatics submitted plans for each of the thousands of parts needed for a rally car. Some were new off-the-shelf parts hijacked from other existing cars, some were custom-designed parts made in several microfactories around the U.S., and some were parts designed to be 3-D printed in any shop. The newest car from Local Motors is a fully 3-D-printed electric car, also designed and manufactured by the community.

Of course, there are many things that are too complex, too unfamiliar, too long term, or too risky to be financed or created by the potential customers. For example, a passenger rocket to Mars, a bridge spanning Alaska and Russia, or a Twitter-based novel are probably out of reach of crowdfunding in the foreseeable future.

But to repeat the lesson from social media: Harnessing the sharing of the crowd will often take you further than you think, and it is almost always the best place to start.

We have barely begun to explore what kinds of amazing things a crowd can do. There must be two million different ways to crowdfund an idea, or to crowdorganize it, or to crowdmake it. There must be a million more new ways to share unexpected things in unexpected ways.

In the next three decades the greatest wealth—and most interesting cultural innovations—lie in this direction. The largest, fastest growing, most profitable companies in 2050 will be companies that will have figured out how to harness aspects of sharing that are invisible and unappreciated today. Anything that can be shared—thoughts, emotions, money, health, time—will be shared in the right conditions, with the right benefits. Anything that can be shared can be shared better, faster, easier, longer, and in a million more ways than we currently realize. At this point in our history, sharing something that has not been shared before, or in a new way, is the surest way to increase its value.

———

In the near future my day will follow a scenario like this: I work as an engineer in a co-op with other engineers from around the world. Our group is collectively owned and managed not by investors, nor by stockholders, but by 1,200 engineers. I earn money for my engineering tweaks. I recently designed a way to improve the efficiency of the flywheel for a regenerative brake on an electric car. If my design is used in the final manufacturing, I get a payment. In fact, anywhere my design is used, even if it is copied for a different car or another purpose, payments still flow back to me automatically. The better the car sells, the higher my micropayments. I'm happy if my work goes viral. The more it is shared, the better. It's the same way photography now works. When I post a photo onto the net, my credentials are encrypted inside the photo image so that the web tracks it and the account of anyone who reposts the photos will pay me a very miniscule micropayment. No matter how many times the picture may be recopied, the credit comes back to me. Compared with last century, it's really easy to make, say, an instructional video now because you can assemble the available parts (images, scenes, even layouts) from other excellent creators, and the micropayments for their work automatically flow back to them as a default. The electric car we are making will be crowdsourced, but unlike decades earlier, every engineer who contributes to the car, no matter how small her contribution, gets paid proportionally.

I have a choice of 10,000 different co-ops I can contribute to. (Not many of my generation want to work for a corporation.) They offer different rates, varying benefits, but, most important, different sets of coworkers. I try to give my favorite co-ops a lot of time not because they pay more, but because I really enjoy working with the best folks—even though we've never met in real life. It is actually hard sometimes to get your work accepted into a high-quality co-op. Your previous contributions—all trackable on the web, of course—have to be really top-notch. They prefer active agents who are contributing to several projects over the years, with multiple streams of automatic payments, as a sign you work well in this sharing economy.

When I am not contributing, I play in a maxed-out virtual world. This world is entirely built by the users—and controlled by them too. I've spent six years constructing this mountaintop village, making every stone wall, every mossy-tiled roof exactly right. I got a lot of cred points for the snow-covered corner, but more important to me is to have it fit perfectly in the greater virtual world we are making. Over 30,000 different games of all types (violent/nonviolent, strategy/shooter) are running on this world platform without interference. In surface area it's almost as big as the moon. There are now 250 million people building the game, each one tending a particular block in this vast world, each one processing on his or her own connected chip. My village runs on my smarthouse monitor. In the past I've lost work to host companies that went out of business, so now I (like millions of others) work only on territory and chips I control. We all contribute our small CPU cycles and storage to the shared Greater World, linked up by a mesh network of rooftop relays. There is a solar-powered mini-relay on my roof that communicates with the other relays on nearby rooftops so that we—the Greater World builders—can't be kicked off a company's network. We collectively run the network, a network no one owns, or rather everyone owns. Our contributions can't be sold, nor do we have to be marketed to while we make and play games within one extended interconnected space. The Greater World is the largest co-op in history, and for the first time we have a hint of a planetary-scale governance. The game world's policies and budget are decided by electronic votes, line by line, facilitated with lots of explaining, tutorials, and even AI. Now over 250 million people want to know why they can't vote on their national budgets that way too.

In a weirdly recursive way, people create teams and co-ops within the Greater World to make stuff in the real world. They find that the tools for collaboration improve quicker in the virtual spaces. I'm contributing to a hackathon that is engineering a collaboratively designed and crowd-funded boomerang probe to Mars, with the goal to be the first to return a few Mars rocks to Earth. Everyone, from geologists to graphic artists, is involved. Just about every high-tech co-op is contributing resources, even man-hours, because they long ago realized the best and newest tools are invented during massively collaborative endeavors like these.

For decades we have been sharing our outputs—our stream of photos, video clips, and well-crafted tweets. In essence, we have been sharing our successes. But only in the last decade did we realize that we learn faster and do better work when we share our failures as well. So in all the collabs I work with, we keep and share all the email, all the chat logs, all correspondence, all intermediate versions, all drafts of everything we do. The entire history is open. We share the process, not just the end product. All the half-baked ideas, dead ends, flops, and redos are actually valuable for both myself and for others hoping to do better. With the entire process out in the open it is harder to fool yourself and easier to see what went right, if it did. Even science has picked up on this idea. When an experiment does not work, scientists are required to share their negative results. I have learned that in collaborative work when you share earlier in the process, the learning and successes come earlier as well. These days I live constantly connected. The bulk of what I share, and what is shared with me, is incremental—constant microupdates, tiny improved versions, minor tweaks—but those steady steps forward feed me. There is no turning the sharing off for long. Even the silence will be shared.

7

FILTERING

There has never been a better time to be a reader, a watcher, a listener, or a participant in human expression. An exhilarating avalanche of new stuff is created every year. Every 12 months we produce 8 million new songs, 2 million new books, 16,000 new films, 30 billion blog posts, 182 billion tweets, 400,000 new products. With little effort today, hardly more than a flick of the wrist, an average person can summon the Library of Everything. You could, if so inclined, read more Greek texts in the original Greek than the most prestigious Greek nobleman of classical times. The same regal ease applies to ancient Chinese scrolls; there are more available to you at home than to emperors of China past. Or Renaissance etchings, or live Mozart concertos, so rare to witness in their time, so accessible now. In every dimension, media today is at an all-time peak of glorious plentitude.

According to the most recent count I could find, the total number of songs that have been recorded on the planet is 180 million. Using standard MP3 compression, the total volume of recorded music for humans would fit onto one 20-terabyte hard disk. Today a 20-terabyte hard disk sells for $2,000. In five years it will sell for $60 and fit into your pocket. Very soon you'll be able to carry around *all* the music of humankind in your pants. On the other hand, if this library is so minuscule, why even bother to carry it around when you could get all music of the world in the cloud streamed to you on demand?

What goes for music also goes for anything and everything that can be rendered in bits. In our lifetime, the entire library of all books, all games, all movies, every text ever printed will be available 24/7 on that same screen thingy or in the same cloud thread. And every day, the library swells. The number of possibilities we confront has been expanded by a growing population, then expanded further by technology that eases creation. There are three times as many people alive today as when I was born (1952). Another billion are due in the next 10 years. An increasing proportion of those extra 5 billion to 6 billion people since my birth have been liberated by the surplus and leisure of modern development to generate new ideas, create new art, make new things. It is 10 times easier today to make a simple video than 10 years ago. It is a hundred times easier to create a small mechanical part and make it real than a century ago. It is a thousand times easier today to write and publish a book than a thousand years ago.

The result is an infinite hall of options. In every direction, countless choices pile up. Despite obsolete occupations like buggy whip maker, the variety of careers to choose from expands. Possible places to vacation, to eat, or even kinds of food all stack up each year. Opportunities to invest explode. Courses to take, things to learn, ways to be entertained explode to astronomical proportions. There is simply not enough time in any lifetime to review the potential of each choice, one by one. It would consume more than a year's worth of our attention to merely preview all the new things that have been invented or created in the previous 24 hours.

The vastness of the Library of Everything quickly overwhelms the very narrow ruts of our own consuming habits. We'll need help to navigate through its wilds. Life is short, and there are too many books to read. Someone, or something, has to choose, or whisper in our ear to help us decide. We need a way to triage. Our only choice is to get assistance in making choices. We employ all manner of filtering to winnow the bewildering spread of options. Many of these filters are traditional and still serve well:

- **We filter by gatekeepers:** Authorities, parents, priests, and teachers shield the bad and selectively pass on "the good stuff."
- **We filter by intermediates:** Sky high is the reject pile in the offices of book publishers, music labels, and movie studios. They say no much more often than yes, performing a filtering function for what gets wide distribution. Every headline in a newspaper is a filter that says yes to this information and ignores the rest.
- **We filter by curators:** Retail stores don't carry everything, museums don't show everything, public libraries don't buy every book. All these curators select their wares and act as filters.
- **We filter by brands:** Faced with a shelf of similar goods, the first-time buyer retreats to a familiar brand because it is a low-effort way to reduce the risk of the purchase. Brands filter through the clutter.
- **We filter by government:** Taboos are prohibited. Hate speech or criticism of leaders or of religion is removed. Nationalistic matters are promoted.
- **We filter by our cultural environment:** Children are fed different messages, different content, different choices depending on the expectations of the schools, family, and society around them.
- **We filter by our friends:** Peers have great sway over our choices. We are very likely to choose what our friends choose.
- **We filter by ourselves:** We make choices based on our own preferences, by our own judgment. Traditionally this is the rarest filter.

None of these methods disappear in the rising superabundance. But to deal with the escalation of options in the coming decades, we'll invent many more types of filtering.

What if you lived in a world where every great movie, book, and song ever produced was at your fingertips as if "for free," and your elaborate system of filters had weeded out the crap, the trash, and anything that would remotely bore you. Forget about all the critically acclaimed creations that mean nothing to you personally. Focus instead on just the things that would truly excite you. Your only choices would be the absolute cream of the cream, the things your best friends would recommend, including a few "random" choices to keep you surprised. In other words, you would encounter only things perfectly matched to you at that moment. You still don't have enough time in your life.

For instance, you could filter your selection of books by reading only the greatest ones. Just focus on the books chosen by experts who have read a lot of them and let them guide you to the 60 volumes considered the best of the very best in Western civilization—the canonical collection known as the Great Books of the Western World. It would take you, or the average reader, some 2,000 hours to completely read all 29 million words. And that's just the Western world. Most of us are going to need further filtering.

The problem is that we start with so many candidates that, even after filtering out all but one in a million, you still have too many. There are more super great five-stars-for-you movies than you can ever watch in your lifetime. There are more useful tools ideally suited to you than you have time to master. There are more cool websites to linger on than you have attention to spare. There are, in fact, more great bands, and books, and gizmos aimed right at you, customized to your unique desires, than you can absorb, even if it was your full-time job.

Nonetheless, we'll try to reduce this abundance to a scale that is satisfying. Let's start with the ideal path. And I'll make it personal. How would I like to choose what I give my attention to next?

First I'd like to be delivered more of what I know I like. This personal filter already exists. It's called a recommendation engine. It is in wide use at Amazon, Netflix, Twitter, LinkedIn, Spotify, Beats, and Pandora, among other aggregators. Twitter uses a recommendation system to suggest who I should follow based on whom I already follow. Pandora uses a similar system to recommend what new music I'll like based on what I already like. Over half of the connections made on LinkedIn arise from their follower recommender. Amazon's recommendation engine is responsible for the well-known banner that "others who like this item also liked this next item." Netflix uses the same to recommend movies for me. Clever algorithms churn through a massive history of everyone's behavior in order to closely predict my own behavior. Their guess is partly based on my own past behavior, so Amazon's banner should really say, "Based on your own history and the history of others similar to you, you should like this." The suggestions are highly tuned to what I have bought and even thought about buying before (they track how long I dwell on a page deliberating, even if I don't choose it). Computing the similarities among a billion past purchases enables their predictions to be remarkably prescient.

These recommendation filters are one of my chief discovery mechanisms. I find them far more reliable, on average, than recommendations from experts or friends. In fact, so many people find these filtered recommendations useful that these kinds of "more like this" offers are responsible for a third of Amazon sales—a difference amounting to about $30 billion in 2014. They are so valuable to Netflix that it has 300 people working on its recommendation system, with a budget of $150 million. There are of course no humans involved in guiding these filters once they are operational. The cognification is based on subtle details of my (and others') behavior that only a sleepless obsessive machine might notice.

The danger of being rewarded with only what you already like, however, is that you can spin into an egotistical spiral, becoming blind to anything slightly different, even if you'd love it. This is called a filter bubble. The technical term is "overfitting." You get stuck at a lower than optimal peak because you behave as if you have arrived at the top, ignoring the adjacent environment. There's a lot of evidence this occurs in the political realm as well: Readers of one political stripe who depend only on a simple filter of "more like this" rarely if ever read books outside their stripe. This overfitting tends to harden their minds. This kind of filter-induced self-reinforcement also occurs in science, the arts, and culture at large. The more effective the "more good stuff like this" filter is, the more important it becomes to alloy it with other types of filters. For instance, some researchers from Yahoo! engineered a way to automatically map one's position in the field of choices visually, to make the bubble visible, which made it easier for someone to climb out of their filter bubble by making small tweaks in certain directions.

Second in the ideal approach, I'd like to know what my friends like that I don't know about. In many ways, Twitter and Facebook serve up this filter. By following your friends, you get effortless updates on the things they find cool enough to share. The ease of shouting out a recommendation via a text or photo is so easy from a phone that we are surprised when someone loves something new but doesn't share it. But friends can also act like a filter bubble if they are too much like you. Close friends can make an echo chamber, amplifying the same choices. Studies show that going to the next circle, to friends of friends, is sometimes enough to enlarge the range of options away from the expected.

A third component in the ideal filter would be a stream that suggested stuff that I don't like but would like to like. It's a bit similar to me trying a least favorite cheese or vegetable every now and then just to see if my tastes have changed. I am sure I don't like opera, but a few years ago I again tried one—*Carmen* at the Met—teleprojected real time in a cinema with prominent subtitles on the huge screen, and I was glad I went. A filter dedicated to probing one's dislikes would have to be delicate, but could also build on the powers of large collaborative databases in the spirit of "people who disliked those, learned to like this one." In somewhat the same vein I also, occasionally, want a bit of stuff I dislike but should learn to like. For me that might be anything related to nutritional supplements, details of political legislation, or hip-hop music. Great teachers have a knack for conveying unsavory packages to the unwilling in a way that does not scare them off; great filters can too. But would anyone sign up for such a filter?

Right now, no one signs up for any of these filters because filters are primarily installed by platforms. The 200 average friends of your average Facebook member already post such a torrent of updates that Facebook feels it must cut, edit, clip, and filter your news to a more manageable stream. You do not see all the posts your friends make. Which ones have been filtered out? By what criteria? Only Facebook knows, and it considers the formulas trade secrets. What it is optimizing for is not even communicated. The company talks about increasing the satisfaction of members, but a fair guess is that it is filtering your news stream to optimize the amount of time you spend on Facebook—a much easier thing to measure than your happiness. But that may not be what you want to optimize Facebook for.

Amazon uses filters to optimize for maximum sales, and that includes filtering the content on the pages you see. Not just what items are recommended, but the other material that appears on the page, including bargains, offers, messages, and suggestions. Like Facebook, Amazon performs thousands of experiments a day, altering their filters to test A over B, trying to personalize the content in response to actual use by millions of customers. They fine-tune the small things, but at such a scale (a hundred thousand subjects at a time) that their results are extremely useful. As a customer I keep returning to Amazon because it is trying to maximize the same thing I am: cheap access to things I will like. That alignment is not always present, but when it is, we return.

Google is the foremost filterer in the world, making all kinds of sophisticated judgments about what search results you see. In addition to filtering the web, it processes 35 billion emails a day, filtering out spam very effectively, assigning labels and priorities. Google is the world's largest collaborative filter, with thousands of interdependent dynamic sieves. If you opt in, it personalizes search results for you and will customize them for your exact location at the time you ask. It uses the now proven principles of collaborative filtering: People who found this answer valuable also found this next one good too (although they don't label it that way). Google filters the content of 60 trillion pages about 2 million times every minute, but we don't often question how it recommends. When I ask it a query, should it show me the most popular, or the most trusted, or the most unique, or the options most likely to please me? I don't know. I say to myself I'd probably like to have the choice to rank results each of those four different ways, but Google knows that all I'd do is look at the first few results and then click. So they say, "Here's the top few we think are the best based on our deep experience in answering 3 billion questions a day." So I click. Google is trying to optimize the chance I'll return to ask it again.

As they mature, filtering systems will be extended to other decentralized systems beyond media, to services like Uber and Airbnb. Your personal preferences in hotel style, status, and service can easily be ported to another system in order to increase your satisfaction when you are matched to a room in Venice. Heavily cognified, incredibly smart filters can be applied to any realm with a lot of choices—which will be more and more realms. Anywhere we want personalization, filtering will follow.

Twenty years ago many pundits anticipated the immediate arrival of large-scale personalization. A 1992 book called *Mass Customization* by Joseph Pine laid out the plan. It seemed reasonable that custom-made work—which was once the purview of the rich—could be widened to the middle class with the right technology. For instance, an ingenious system of digital scans and robotic flexible manufacturing could provide personally tailored shirts for the middle class, instead of just bespoke shirts for the gentry. A few startups tried to execute "mass customization" for jeans, shirts, and baby dolls in the late 1990s, but they failed to catch on. The main hurdle was that, except in trivial ways (choosing a color or length), it was very difficult to capture or produce significant uniqueness without raising prices to the luxury level. The vision was too far ahead of the technology. But now the technology is catching up. The latest generation of robots are capable of agile manufacturing, and advanced 3-D printers can rapidly produce units of one. Ubiquitous tracking, interacting, and filtering means that we can cheaply assemble a multidimensional profile of ourselves, which can guide any custom services we desire.

Here is a picture of where this force is taking us. My day in the near future will entail routines like this: I have a pill-making machine in my kitchen, a bit smaller than a toaster. It stores dozens of tiny bottles inside, each containing a prescribed medicine or supplement in powdered form. Every day the machine mixes the right doses of all the powders and stuffs them all into a single personalized pill (or two), which I take. During the day my biological vitals are tracked with wearable sensors so that the effect of the medicine is measured hourly and then sent to the cloud for analysis. The next day the dosage of the medicines is adjusted based on the past 24-hour results and a new personalized pill produced. Repeat every day thereafter. This appliance, manufactured in the millions, produces mass personalized medicine.

My personal avatar is stored online, accessible to any retailer. It holds the exact measurements of every part and curve of my body. Even if I go to a physical retail store, I still try on each item in a virtual dressing room before I go because stores carry only the most basic colors and designs. With the virtual mirror I get a surprisingly realistic preview of what the clothes will look like on me; in fact, because I can spin my simulated dressed self around, it is more revealing than a real mirror in a dressing room. (It could be better in predicting how comfortable the new clothes feel, though.) My clothing is custom fit based on the specifications (tweaked over time) from my avatar. My clothing service generates new variations of styles based on what I've worn in the past, or on what I spend the most time wishfully gazing at, or on what my closest friends have worn. It is filtering styles. Over years I have trained an in-depth profile of my behavior, which I can apply to anything I desire.

My profile, like my avatar, is managed by Universal You. It knows that I like to book inexpensive hostels when I travel on vacation, but with a private bath, maximum bandwidth, and always in the oldest part of the town, except if it is near a bus station. It works with an AI to match, schedule, and reserve the best rates. It is more than a mere stored profile; rather it is an ongoing filter that is constantly adapting to wherever I have already gone, what kind of snapshots and tweets I made about past visits, and it weighs my new interests in reading and movies since books and movies are often a source for travel desires. It pays a lot of attention to the travels of my best friends and their friends, and from that large pool of data often suggests specific restaurants and hostels to visit. I generally am delighted by its recommendations.

Because my friends let Universal You track their shopping, eating out, club attendance, movie streaming, news screening, exercise routines, and weekend excursions, it can make very detailed recommendations for me—with minimal effort on their part. When I wake in the morning, Universal filters through my update stream to deliver the most vital news of the type I like in the morning. It filters based on the kinds of things I usually forward to others, or bookmark, or reply to. In my cupboard I find a new kind of cereal with saturated nutrition that my friends are trying this week, so Universal ordered it for me yesterday. It's not bad. My car service notices where the traffic jams are this morning, so it schedules my car later than normal and it will try an unconventional route to the place I'll work today, based on several colleagues' commutes earlier. I never know for sure where my office will be since our startup meets in whatever coworking space is available that day. My personal device turns the space's screens into my screen. My work during the day entails tweaking several AIs that match doctoring and health styles with clients. My job is to help the AIs understand some of the outlier cases (such as folks with faith-healing tendencies) in order to increase the effectiveness of the AIs' diagnoses and recommendations.

When I get home, I really look forward to watching the string of amusing 3-D videos and fun games that Albert lines up for me. That's the name I gave to the avatar from Universal who filters my media for me. Albert always gets the coolest stuff because I've trained him really well. Ever since high school I would spend at least 10 minutes every day correcting his selections and adding obscure influences, really tuning the filters, so that by now, with all the new AI algos and the friends of friends of friends' scores, I have the most amazing channel. I have a lot of people who follow my Albert daily. I am at the top of the leaderboard for the VR worlds filter. My mix is so popular that I'm earning some money from Universal—well, at least enough to pay for all my subscriptions.

———

We are still at the early stages in how and what we filter. These powerful computational technologies can be—and will be—applied to the internet of everything. The most trivial product or service could be personalized if we wanted it (but many times we won't). In the next 30 years the entire cloud will be filtered, elevating the degree of personalization.

Yet every filter throws something good away. Filtering is a type of censoring, and vice versa. Governments can implement nationwide filters to remove unwanted political ideas and restrict speech. Like Facebook or Google, they usually don't disclose what they are filtering. Unlike social media, citizens don't have an alternative government to switch to.

But even in benign filtering, by design we see only a tiny fraction of all there is to see. This is the curse of the postscarcity world: We can connect to only a thin thread of all there is. Each day maker-friendly technologies such as 3-D printing, phone-based apps, and cloud services widen the sky of possibilities another few degrees. So each day wider filters are needed to access this abundance at human scale. There is no retreat from more filtering. The inadequacies of a filter cannot be remedied by eliminating filters. The inadequacies of a filter can be remedied only by applying countervailing filters upon it.

From the human point of view, a filter focuses content. But seen in reverse, from the content point of view, a filter focuses human attention. The more content expands, the more focused that attention needs to become. Way back in 1971 Herbert Simon, a Nobel Prize–winning social scientist, observed, "In an information-rich world, the wealth of information means a dearth of something else: a scarcity of whatever it is that information consumes. What information consumes is rather obvious: it consumes the attention of its recipients. Hence a wealth of information creates a poverty of attention." Simon's insight is often reduced to "In a world of abundance, the only scarcity is human attention."

Our attention is the only valuable resource we personally produce without training. It is in short supply and everyone wants some of it. You can stop sleeping altogether and you will still have only 24 hours per day of potential attention. Absolutely nothing—no money or technology—will ever increase that amount. The maximum potential attention is therefore fixed. Its production is inherently limited while everything else is becoming abundant. Since it is the last scarcity, wherever attention flows, money will follow.

Yet for being so precious, our attention is relatively inexpensive. It is cheap, in part, because we have to give it away each day. We can't save it up or hoard it. We have to spend it second by second, in real time.

In the United States, TV still captures most of our attention, followed by radio, and then the internet. These three take the majority of our attention, while the others—books, newspapers, magazines, music, home video, games—consume only slivers of the total pie.

But not all attention is equal. In the advertising business, quantity of attention is often reflected in a metric called CPM, or cost per thousand (*M* is Latin for "thousand"). That's a thousand views, or a thousand readers or listeners. The estimated average CPM of various media platforms ranges widely. Cheap outdoor billboards average $3.50, TV is $7, magazines earn $14, and newspapers $32.50.

There's another way to calculate how much our attention is worth. We can tally up the total annual revenue earned by each of the major media industries, and the total amount of time spent on each media, and then calculate how much revenue each hour of attention generates in dollars per hour. The answer surprised me.

First, it is a low number. The ratio of dollars earned by the industry per hour of attention spent by consumers shows that attention is not worth very much to media businesses. While half a trillion hours are devoted to TV annually (just in the U.S.), it generates for its content owners, on average, only 20 cents per hour. If you were being paid to watch TV at this rate, you would be earning a third-world hourly wage. Television watching is coolie labor. Newspapers occupy a smaller slice of our attention, but generate more revenue per hour spent with them—about 93 cents per hour. The internet, remarkably, is relatively more expensive, increasing its quality of attention each year, garnering on average $3.60 per hour of attention.

A lousy 20 cents per hour of attention that we watchers "earn" for TV companies, or even a dollar an hour for upscale newspapers, reflects the worth of what I call "commodity attention." The kind of attention we pay to entertainment commodities that are easily duplicated, easily transmitted, nearly ubiquitous, and always on is not worth much. When we inspect how much we have to pay to purchase commodity content—all the content that can easily be copied—such as books, movies, music, news, etc.—the rates are higher, but still don't reflect the fact that our attention is the last scarcity. Take a book, for instance. The average hardcover book takes 4.3 hours to read and $23 to buy. Therefore the average consumer cost for that reading duration is $5.34 per hour. A music CD is, on average, listened to dozens of times over its lifetime, so its retail price is divided by its total listening time to arrive at its hourly rate. A two-hour movie in a theater is seen only once, so its per hour rate is half the ticket price. These rates can be thought of as mirroring how much we, as the audience, value our attention.

In 1995 I calculated the average hourly costs for various media platforms, including music, books, newspapers, and movies. There was some variation between media, but the price stayed within the same order of magnitude, converging on a mean of $2.00 per hour. In 1995 we tended to pay, on average, two bucks per hour for media use.

Fifteen years later, in 2010, and then again in 2015, I recalculated the values for a similar set of media using the same method. When I adjusted for inflation and translated into 2015 dollars, the average cost to consume one hour of media in 1995, 2010, and 2015 is respectively $3.08, $2.69, and $3.37. That means that the value of our attention has been remarkably stable over 20 years. It seems we have some intuitive sense of what a media experience "should" cost, and we don't stray much from that. It also means that companies making money from our attention (such as many high-profile tech companies) are earning only an average of $3 per hour of attention—if they include high-quality content.

In the coming two decades the challenge and opportunity is to harness filtering technologies to cultivate higher quality attention at scale. Today, the bulk of the internet economy is fueled by trillions of hours of low-grade commodity attention. A single hour by itself is not worth much, but en masse it can move mountains. Commodity attention is like a wind or an ocean tide: a diffuse force that must be captured with large instruments.

————

The brilliance behind Google, Facebook, and other internet platforms' immense prosperity is a massive infrastructure that filters this commodity attention. Platforms use serious computational power to match the expanding universe of advertisers to the expanding universe of consumers. Their AIs seek the optimal ad at the optimal time in the optimal place and the optimal frequency with the optimal way to respond. While this is sometimes termed personalized advertising, it is in fact far more complex than just targeting ads to individuals. It represents an ecosystem of filterings, which have consequences beyond just advertising.

Anyone can sign up to be an advertiser on Google by filling out an online form. (Most of the ads are text, like a classified ad.) That means the number of potential advertisers might be in the billions. You could be a small-time businessperson advertising a cookbook for vegan backpackers or a new baseball glove you invented. On the other side of the equation, anyone running a web page for any reason can allow an advertiser to place an ad on their page and potentially earn income from this advertising. The web page could be a personal blog or a company home page. For about eight years I ran Google AdSense ads on my own personal blogs. The hundred dollars or so I earned each month for showing ads was small potatoes for a billion-dollar company, but the tiny size of these transactions didn't matter to Google because it was all automated, and the tiny sums would add up. The AdSense network embraces all comers no matter how small, so the potential places an ad could run swells to the billions. To mathematically match these billions of possibilities—of billions of people wanting to advertise and billions of places willing to run ads—an astronomical number of potential solutions are needed. In addition, the optimal solutions can shift by time of day or geographical location—and so Google (and other search companies like Microsoft and Yahoo!) need their gigantic cloud computers to sort through them.

To match advertiser with reader, Google's computers roam the web 24 hours a day and collect all the content on every one of the 60 trillion pages on the web and store that information in its huge database. That's how Google delivers you an instant answer whenever you query it. It has already indexed the location of every word, phrase, and fact on the web. So when a web owner wants to allow a small AdSense ad to run on their blog page, Google summons up its record of what material is on that page and then uses its superbrain to find someone—right that minute—who wants to place an ad related to that material. When the match consummates, the ad on the web page will reflect the editorial content of the page. Suppose the website belongs to a small-town softball team; the ads for an innovative baseball mitt would be very appropriate for that context. Readers are much more likely to click on it than an ad for snorkeling gear. So Google, guided by the context of the material, will place mitt ads on softball websites.

But that's just the start of the complexity, because Google will try to make it a three-way match. Ideally, the ads not only match the context of the web page, but also the interest of the reader visiting the page. If you arrive at a general news site—say, CNN—and it knows you play in a softball league, you might see more ads for sports equipment than for furniture. How does it know about you? Unbeknownst to most people, when you arrive at a website you arrive with a bunch of invisible signs hanging around your neck that display where you just came from. These signs (technically called cookies) can be read not just by the website you have arrived at, but by many of the large platforms—like Google—who have their fingers all over the web. Since almost every commercial website uses a Google product, Google is able to track your journey as you visit one page after another all across the web. And of course if you google anything, it can follow you from there as well. Google does not know your name, address, or email (yet), but it does remember your web behavior. So if you arrive at a news site after visiting a softball team page, or after googling "softball mitt," it can make some assumptions. It takes these guesses and adds them to the calculation of figuring out what ads to place on a web page that you've just arrived at. It's almost magical, but the ads you see on a website today are not added until the moment after you land there. So in real time Google and the news site will select the ad that you see, so that you see a different ad than I would. If the whole ecosystem of filters is working, the ad you see will reflect your recent web visit history and will incline more to your interests.

But wait—there's more! Google itself becomes a fourth party in this multisided market. In addition to satisfying the advertisers, the web page publisher, and the reader, Google is also trying to optimize its own score. Some audiences' attention is worth more to advertisers than others. Readers of health-related websites are valuable because they may potentially spend a lot of money on pills and treatments over a long period of time, whereas readers of a walking club forum buy shoes only once in a while. So behind each placement is a very complicated auction that matches the value of key context words ("asthma" will cost a lot more than "walking") with the price an advertiser is willing to pay *along with* the performance level of readers who actually click on the ad. The advertiser pays a few cents to the web page owner (and to Google) if someone clicks on the ad, so the algorithms try to optimize the placement of the ads, the rates that are charged, and the rate they are engaged. A 5-cent ad for a softball glove that gets clicked 12 times is worth more than a 65-cent ad for an asthma inhaler that gets clicked once. But then the next day the softball team blog posts a warning about the heavy pollen count this spring, and suddenly advertising inhalers on the softball blog is worth 85 cents. Google may have to juggle hundreds of millions of factors all at once, in real time, in order to settle on the optimal arrangement for that hour. When everything works in this very fluid four-part match, Google's income is also optimized. In 2014, 21 percent of Google's total revenue, or $14 billion, came through this system of AdSense ads.

This complicated zoo of different types of interacting attention was nearly unthinkable before the year 2000. The degree of cognification and computation required to track, sort, and filter each vector was beyond practical. But as systems of tracking and cognifying and filtering keep growing, ever more possible ways to arrange attention—both giving and receiving—are made feasible. This period is analogous to the Cambrian era of evolution, when life was newly multicellular. In a very brief period (geologically speaking), life incarnated many previously untried possibilities. It racked up so many new, and sometimes strange, living arrangements so fast that we call this historical period of biological innovation the Cambrian explosion. We are at a threshold of a Cambrian explosion in attention technology, as novel and outlandish versions of attention and filtering are given a try.

For instance, what if advertising followed the same trend of decentralization as other commercial sectors have? What if customers created, placed, and paid for ads?

Here is one way to think of this strange arrangement. Each enterprise that is supported by advertising—which is currently the majority of internet companies—needs to convince advertisers to place their ads with them in particular. The argument a publisher, conference, blog, or platform makes to companies is that no one else can reach the particular audience they reach, or reach them within as good a relationship. The advertisers have the money, so they are picky about who gets to run their ads. While a publication will try to persuade the most desirable advertisers, the publications don't get to select which ads run. The advertisers, or their agents, do. A magazine fat with ads or a TV show crammed with commercials usually considers itself lucky to have been picked as the vehicle for the ads.

But what if anyone with an audience could choose the particular ads they wanted to display, without having to ask permission? Say you saw a really cool commercial for a running shoe and you wanted to include it in your stream—and get paid for it just as a TV station would. What if any platform could simply gather the best ads that appealed to them and then were paid for the ones they ran—and were watched—according to the quality and quantity of traffic brought to them? Ads that were videos, still images, audio files would contain embedded codes that kept track of where they were shown and how often they were viewed, so that no matter how often they were copied, the host at the time would get paid. The very best thing that can happen to an ad is that it goes viral, getting placed and replayed on as many platforms as possible. Because an ad played on your site might generate some revenue for your site, you'll be on the lookout for memorable ads to host. Imagine a Pinterest board that collected ads. Any ad in the collection that was played or viewed by readers would generate revenue for the collector. If done well, the audience might come not only for cool content but for cool ads—in the way millions of people show up for the Super Bowl on TV in large part to watch the commercials.

The result would create a platform that curated ads as well as content. Editors would spend as much time hunting down unknown, little-seen, attention-focusing ads as they might spend on finding news articles. However, wildly popular ads may not pay as much as niche ads. Obnoxious ads might pay more than humorous ones. So there will be a trade-off between cool-looking ads that make no money versus square but profitable ones. And of course, fun, high-paying ads would be likely shown a lot, both decreasing their coolness and probably decreasing their price. There might be magazines/publications/online websites that contained nothing but artfully arranged ads—and they would make money. There are websites today that feature only movie trailers or great commercials, but they don't earn anything from the sources for hosting them. Soon enough they will.

This arrangement completely reverses the power of the established ad industry. Like Uber and other decentralized systems, it takes what was once a highly refined job performed by a few professionals and spreads it across a peer-to-peer network of amateurs. No advertising professional in 2016 believes it could work, and even reasonable people think it sounds crazy, but one thing we know about the last 30 years is that seemingly impossible things can be accomplished by peers of amateurs when connected smartly.

A couple of maverick startups in 2016 are trying to disrupt the current attention system, but it may take a number of tries before some of the radical new modes stick. The missing piece between this fantasy and reality is the technology to track the visits, to weed out fraud, and quantify the attention that a replicating ad gets, and then to exchange this data securely in order to make a correct payment. This is a computational job for a large multisided platform such as Google or Facebook. It would require a lot of regulation because the money would attract fraudsters and creative spammers. But once the system was up and running, advertisers would release ads to virally zip around the web. You catch one and embed it in a site. It then triggers a payment if a reader clicks on it.

This new regime puts the advertisers in a unique position. Ad creators no longer control where an ad will show up. This uncertainty would need to be compensated in some way by the ad's construction. Some would be designed to replicate quickly and to induce action (purchases) by the viewers. Other ads may be designed to sit monumentally where they are, not travel, and to slowly affect branding. Since an ad could, in theory, be used like an editorial, it might resemble editorial material. Not all ads would be released into the wilds. Some, if not many, ads might be saved for traditional directed placement (making them rare). The success of this system would only prosper in addition to, and layered on top of, the traditional advertising modes.

The tide of decentralization floods every corner. If amateurs can place ads, why can't the customers and fans create the ads themselves? Technology may be able to support a peer-to-peer ad *creation* network.

A couple of companies have experimented with limited versions of user-created ads. Doritos solicited customer-generated video commercials to be aired on the 2006 Super Bowl. It received 2,000 video ads and more than 2 million people voted on the best, which was aired. Every year since then it has received on average 5,000 user-made submissions. Doritos now awards $1 million to the winner, which is far less than what professional ads cost. In 2006, GM solicited user-created ads for its Chevy Tahoe SUV and received 21,000 of them (4,000 were negative ads complaining about SUVs). These examples are limited because the only ads that ran had to be approved and processed through company headquarters, which is not truly peer to peer.

A fully decentralized peer-to-peer user-generated crowdsourced ad network would let users create ads, and then let user-publishers choose which ads they wanted to place on their site. Those user-generated ads that actually produced clicks would be kept and/or shared. Those that weren't effective would be dropped. Users would become ad agencies, as they have become everything else. Just as there are amateurs making their living shooting stock photos or working tiny spreads on eBay auctions, there will surely be many folks who will earn a living churning out endless variations of ads for mortgages.

I mean, really, who would you rather make your ads? Would you rather employ the expensive studio pros who come up with a single campaign using their best guess, or a thousand creative kids endlessly tweaking and testing their ads of your product? As always, it will be a dilemma for the crowd: Should they work on an ad for a reliable bestseller—and try to better a thousand others with the same idea—or go for the long tail, where you might have an unknown product all to yourself if you get it right? Fans of products would love to create ads for it. Naturally they believe no one else knows it as well as they do, and that the current ads (if any) are lame, so they will be confident and willing to do a better job.

How realistic is it to expect big companies to let go of their advertising? Not very. Big companies are not going to be the first to do this. It will take many years of brash upstarts with small to no advertising budgets who have little to lose figuring this out. As with AdSense, big is not where the leverage is. Rather this new corner of ad space liberates the small to middle—a billion businesses who would have never thought of, let alone ever got around to, developing a cool advertising campaign. With a peer-to-peer system, these ads would be created by passionate (and greedy) users and unleashed virally into the blog wilds, where the best ads would evolve by testing and redesign until they were effective.

By tracing alternative routes of attention, we can see that there are many yet untapped formations of attention. Esther Dyson, an early internet pioneer and investor, has long complained of the asymmetry of attention in email. Since she has been active in forming the governance of the internet and financing many innovative startups, her inbox overflows with mail from people she doesn't know. She says, "Email is a system that lets other people add things to my to-do list." Right now there is no cost for adding an email in someone else's queue. Twenty years ago she proposed a system that would enable someone to charge senders for reading their email. In other words, you'd have to pay Esther to read your email to her. She might charge as little as 25 cents for some senders—say, students—or more (say, $2) for a press release from a PR company. Friends and family are probably not charged, but a complicated pitch from an entrepreneur might warrant a $5 fee. Charges can also be forgiven retroactively once a piece of mail is read. Of course, Esther is a sought-after investor, so her default filter may be set high—say, $3 per email message she reads. An average person won't command the same fee, but *any* charge acts as a filter. More important, a sufficient fee to read acts as a signal to the recipient that the message is deemed "important."

The recipient doesn't need to be as famous as Esther to be worth paying to read an email. It could involve a small-time influencer. An extremely powerful use of the cloud is to untangle the tangled network of followers and followed. Massive cognification can trace out every per-mutation of who is influencing whom. People who influence a small number of people who in turn influence others may get a different rank-ing than people who influence a whole lot of people who don't influence others. Status is very local and specific. A teenage girl with a lot of loyal friends who follow her lead in fashion could have a much higher influence rank than a CEO of a tech company. This relationship network analysis can go to the third and fourth level (the friend of a friend of a friend) in an explosion of computational complexity. Out of this complexity vari-ous types of scores can be assigned for degrees of influence and attention. A high scorer may charge more to read an email, but may also choose to adjust what is charged based on the scores of the sender—which adds further complexity and costs to calculating the sum.

The principle of paying people directly for their attention can be extended to advertising as well. We spend our attention on ads for free. Why don't we charge companies to watch their commercials? As in Esther's scheme, different people might charge different fees depending on the source of the ad. And different people would have different desir-ability quotients for the vendors. Some watchers would be worth a lot. Retailers speak about the total lifetime spending of a customer; a cus-tomer predicted to spent $10,000 over his or her lifetime at a particular retailer's store would be worth an early $200 discount bonus. There might also be a total lifetime influence for customers as well, as their influence ripples out to the followers of followers of followers, and so on. The sum could be tallied up and estimated for their lifespan. For those attention-givers with a high estimated lifetime influence, a company might find it worthwhile to pay them directly instead of paying advertisers. The com-pany could pay in either cash or valuable goods and services. This is essentially what the swag bags given away at the Oscar Award ceremonies do. In 2015 the bags for some nominees were crammed with $168,000 worth of merchandise, a mixture of consumer commodities like lip gloss, lollipops, travel pillows, and luxury hotel and travel packages. Vendors make the reasonable calculation that Oscar nominees are high influencers. The recipients don't need any of this stuff, but they might gab about their gifts to their fans.

The Oscars are obviously an outlier. But on a smaller scale, locally well-known people can gather a significantly loyal following and earn a sizable lifetime influence score. But until recently it was impossible to pinpoint the myriad microcelebrities in a population of hundreds of millions. Today, advances in filtering technology and sharing media enable these mavens to be spotted and reached in bulk. Instead of the Oscars, retailers can aim at a huge network of smaller influencers. Companies that normally advertise could skip ads altogether. They would take their million-dollar advertising budgets and directly pay the accounts of tens of thousands of small-time influencers for their attention.

We have not yet explored all the possible ways to exchange and manage attention and influence. A blank continent is opening up. Many of the most interesting possible modes—like getting paid for your attention or influence—are still unborn. The future forms of attention will emerge from a choreography of streams of influence that are subject to tracking, filtering, sharing, and remixing. The scale of data needed to orchestrate this dance of attention reaches new heights of complexity.

Our lives are already significantly more complex than even five years ago. We need to pay attention to far more sources in order to do our jobs, to learn, to parent, or even to be entertained. The number of factors and possibilities we have to attend to rises each year almost exponentially. Thus our seemingly permanently distracted state and our endless flitting from one thing to another is not a sign of disaster, but is a necessary adaptation to this current environment. Google is not making us dumber. Rather we need to web surf to be agile, to remain alert to the next new thing. Our brains were not evolved to deal with zillions. This realm is beyond our natural capabilities, and so we have to rely on our machines to interface with it. We need a real-time system of filters upon filters in order to operate in the explosion of options we have created.

———

A major accelerant in this explosion of superabundance—the superabundance that demands constant increases in filtering—is the compounding cheapness of stuff. In general, on average, over time technology tends toward the free. That tends to make things abundant. At first it may be hard to believe that technology wants to be free. But it's true about most things we make. Over time, if a technology persists long enough, its costs begin to approach (but never reach) zero. In the goodness of time any particular technological function will act as if it were free. This slide toward the free seems to be true for basic things like foodstuffs and materials (often called commodities), and complicated stuff like appliances, as well as services and intangibles. The costs of all these (per fixed unit) has been dropping over time, particularly since the industrial revolution. According to a 2002 paper published by the International Monetary Fund, "There has been a downward trend in real commodity prices of about 1 percent per year over the last 140 years." For a century and a half prices have been headed toward zero.

This is not just about computer chips and high-tech gear. Just about everything we make, in every industry, is headed in the same economic direction, getting cheaper every day. Let's take just one example: the dropping cost of copper. Plotted over the long term (since 1800), the graph of its price drifts downward. While it trends toward zero (despite ups and downs), the price will never reach its limit of the absolutely free. Instead it steadily creeps closer and closer to the ideal limit, in an infinite series of narrowing gaps. This pattern of paralleling the limit but never crossing it is called approaching the asymptote. The price here is not zero, but effectively zero. In the vernacular it is known as "too cheap to meter"—too close to zero to even keep track of.

That leaves the big question in an age of cheap plentitude: What is really valuable? Paradoxically, our attention to commodities is not worth much. Our monkey mind is cheaply hijacked. The remaining scarcity in an abundant society is the type of attention that is not derived or focused on commodities. The only things that are increasing in cost while everything else heads to zero are human experiences—which cannot be copied. Everything else becomes commoditized and filterable.

The value of experience is rising. Luxury entertainment is increasing 6.5 percent annually. Spending at restaurants and bars increased 9 percent in 2015 alone. The price of the average concert ticket has increased by nearly 400 percent from 1981 to 2012. Ditto for the price of health care in the United States. It rose 400 percent from 1982 to 2014. The average U.S. rate for babysitting is $15 per hour, twice the minimum wage. In big U.S. cities it is not unusual for parents to spend $100 for child care during an evening out. Personal coaches dispensing intensely personal attention for a very bodily experience are among the fastest growing occupations. In hospice care, the cost of drugs and treatments is in decline, but the cost of home visits—experiential—is rising. The cost of weddings has no limit. These are not commodities. They are experiences. We give them our precious, scarce, fully unalloyed attention. To the creators of these experiences, our attention is worth a lot. Not coincidentally, humans excel at creating and consuming experiences. This is no place for robots. If you want a glimpse of what we humans do when the robots take our current jobs, look at experiences. That's where we'll spend our money (because they won't be free) and that's where we'll make our money. We'll use technology to produce commodities, and we'll make experiences in order to avoid becoming a commodity ourselves.

The funny thing about a whole class of technology that enhances experience and personalization is that it puts great pressure on us to know who we are. We will soon dwell smack in the middle of the Library of Everything, surrounded by the liquid presence of all existing works of humankind, just within reach of our fingertips, for free. The great filters will be standing by, quietly guiding us, ready to serve us our wishes. "What do you want?" the filters ask. "You can choose anything; what do you choose?" The filters have been watching us for years; they anticipate what we will ask. They can almost autocomplete it right now. Thing is, we don't know what we want. We don't know ourselves very well. To some degree we will rely on the filters to tell us what we want. Not as slave masters, but as a mirror. We'll listen to the suggestions and recommendations that are generated by our own behavior in order to hear, to see who we are. The hundred million lines of code running on the million servers of the intercloud are filtering, filtering, filtering, helping us to distill ourselves to a unique point, to optimize our personality. The fears that technology makes us more uniform, more commoditized are incorrect. The more we are personalized, the easier it is for the filters because we become distinct, an actualized distinction they can reckon with. At its heart, the modern economy runs on distinction and the power of differences—which can be accentuated by filters and technology. We can use the mass filtering that is coming to sharpen who we are, for the personalization of our own person.

More filtering is inevitable because we can't stop making new things. Chief among the new things we will make are new ways to filter and personalize, to make us more like ourselves.

8

REMIXING

Paul Romer, an economist at New York University who specializes in the theory of economic growth, says real sustainable economic growth does not stem from new resources but from existing resources that are rearranged to make them more valuable. Growth comes from remixing. Brian Arthur, an economist at the Santa Fe Institute who specializes in the dynamics of technological growth, says that all new technologies derive from a combination of existing technologies. Modern technologies are combinations of earlier primitive technologies that have been rearranged and remixed. Since one can combine hundreds of simpler technologies with hundreds of thousands of more complex technologies, there is an unlimited number of possible new technologies—but they are all remixes. What is true for economic and technological growth is also true for digital growth. We are in a period of productive remixing. Innovators recombine simple earlier media genres with later complex genres to produce an unlimited number of new media genres. The more new genres, the more possible newer ones can be remixed from them. The rate of possible combinations grows exponentially, expanding the culture and the economy.

We live in a golden age of new mediums. In the last several decades hundreds of media genres have been born, remixed out of old genres. Former mediums such as a newspaper article, or a 30-minute TV sitcom, or a 4-minute pop song still persist and enjoy immense popularity. But digital technology unbundles those forms into their elements so they can be recombined in new ways. Recent newborn forms include a web list article (a listicle) or a 140-character tweet storm. Some of these recombined forms are now so robust that they serve as a new genre. These new genres themselves will be remixed, unbundled, and recombined into hundreds of other new genres in the coming decades. Some are already mainstream—they encompass at least a million creators, and hundreds of millions in their audience.

For instance, behind every bestselling book are legions of fans who write their own sequels using their favorite author's characters in slightly altered worlds. These extremely imaginative extended narratives are called fan fiction, or fanfic. They are unofficial—without the original authors' cooperation or approval—and may mix elements from more than one book or author. Their chief audience is other avid fans. One fanfic archive lists 1.5 million fan-created works to date.

Extremely short snips (six seconds or less) of video quickly recorded on a phone can easily be shared and reshared with an app called Vine. Six seconds is enough for a joke or a disaster to spread virally. These brief recorded snips may be highly edited for maximum effect. Compilations of a sequence of six-second vines are a popular viewing mode. In 2013, 12 million Vine clips were posted to Twitter every day, and in 2015 viewers racked up 1.5 billion daily loops. There are stars on Vine with a million followers. But there is another kind of video that is even shorter. An animated gif is a seemingly still graphic that loops through its small motion again and again and again. The cycle lasts only a second or two, so it could be thought of as a one-second video. Any gesture can be looped. A gif might be a quirky expression on a face that is repeated, or a famous scene from a movie put on a loop, or it could be a repeating pattern. The endless repetition encourages it to be studied closely until it transcends into something bigger. Of course, there are entire websites devoted to promoting gifs.

These examples can only hint at the outburst and sheer frenzy of new forms appearing in the coming decades. Take any one of these genres and multiply it. Then marry and crossbreed them. We can see the nascent outlines of the new ones that might emerge. With our fingers we will drag objects out of films and remix them into our own photos. A click of our phone camera will capture a landscape, then display its history in words, which we can use to annotate the image. Text, sound, motion will continue to merge. With the coming new tools we'll be able to create our visions on demand. It will take only a few seconds to generate a believable image of a turquoise rose, glistening with dew, poised in a trim golden vase—perhaps even faster than we could write these words. And that is just the opening scene.

The supreme fungibility of digital bits allows forms to morph easily, to mutate and hybridize. The quick flow of bits permits one program to emulate another. To simulate another form is a native function of digital media. There's no retreat from this multiplicity. The number of media choices will only increase. The variety of genres and subgenres will continue to explode. Sure, some will rise in popularity while others wane, but few will disappear entirely. There will still be opera lovers a century from now. But there will be a billion video game fans and a hundred million virtual reality worlds.

The accelerating fluidity of bits will continue to overtake media for the next 30 years, furthering a great remixing.

———

At the same time, the cheap and universal tools of creation (megapixel phone cameras, YouTube Capture, iMovie) are quickly reducing the effort needed to create moving images and upsetting a great asymmetry that has been inherent in all media. That is: It is easier to read a book than to write one, easier to listen to a song than to compose one, easier to attend a play than to produce one. Feature-length classic movies in particular have long suffered from this user asymmetry. The intensely collaborative work needed to coddle pieces of chemically treated film and paste them together into movies meant that it was *vastly* easier to watch a movie than to make one. A Hollywood blockbuster can take a million person-hours to produce and only two hours to consume. To the utter bafflement of the experts who confidently claimed that viewers would never rise from their reclining passivity, tens of millions of people have in recent years spent uncountable hours making movies of their own design. Having a ready and reachable audience of potential billions helps, as does the choice of multiple modes in which to create. Because of new consumer gadgets, community training, peer encouragement, and fiendishly clever software, the ease of making video now approaches the ease of writing.

This is not how Hollywood makes films, of course. A blockbuster film is a gigantic creature custom built by hand. Like a Siberian tiger, it demands our attention—but it is also very rare. Every year about 600 feature films are released in North America, or about 1,200 hours of moving images. As a percentage of the hundreds of millions of hours of moving images produced annually today, 1,200 hours is minuscule. It is an insignificant rounding error.

We tend to think the tiger represents the animal kingdom, but in truth a grasshopper is a truer statistical example of an animal. The handcrafted Hollywood film is a rare tiger. It won't go away, but if we want to see the future of motion pictures, we need to study the swarming critters below—the jungle of YouTube, indie films, TV serials, documentaries, commericals, infomercials, and insect-scale supercuts and mashups—and not just the tiny apex of tigers. YouTube videos are viewed more than 12 billion times in a single month. The most viewed videos have been watched several billion times each, more than any blockbuster movie. More than 100 million short video clips with very small audiences are shared to the net every day. Judged merely by volume and the amount of attention the videos collectively garner, these clips are now the center of our culture. Their craftsmanship varies widely. Some are made with the same glossiness as a Hollywood movie, but most are made by kids in their kitchen with a phone. If Hollywood is at the apex of the pyramid, the bottom is where the swampy action is, and where the future of the moving image begins.

The vast majority of these non-Hollywood productions rely on remixing, because remixing makes it much easier to create. Amateurs take soundtracks found online, or recorded in their bedrooms, cut and reorder scenes, enter text, and then layer in a new story or novel point of view. Remixing of commercials is rampant. Each genre often follows a set format.

For example, remixed movie trailers. Movie trailers themselves are a recent art form. Because of their brevity and compact storytelling, movie trailers can be easily recut into alternative narratives—for instance, movie trailers for imaginary movies. An unknown amateur may turn a comedy into a horror flick, or vice versa. Remixing the soundtrack of the trailer is a common way to mash up these short movies. Some fans create music videos made by matching and mixing a pop song soundtrack with edited clips from obscure cult hit movies. Or they clip scenes from a favorite movie or movie star, which are then edited to fit an unlikely song. These become music videos for a fantasy universe. Rabid fans of pop bands will take their favorite songs on video and visibly add the song's lyrics in large type. Eventually these lyric videos became so popular that some bands started releasing official music videos with lyrics. As the words float over visuals in sync with the sounds, this is a true remixing and convergence of text and image—video you read, music you watch.

Remixing video can even become a kind of collective sport. Hundreds of thousands of passionate anime fans around the world (meeting online, of course) remix Japanese animated cartoons. They clip the cartoons into tiny pieces, some only a few frames long, then rearrange them with video editing software and give them new soundtracks and music, often with English dialogue. This probably involves far more work than was required to draw the original cartoon, but far less work than it would have required to create a simple clip 30 years ago. The new anime vids tell completely new stories. The real achievement in this subculture is to win the Iron Editor challenge. Just as in the TV cookoff contest *Iron Chef*, the Iron Editor must remix videos in real time in front of an audience while competing with other editors to demonstrate superior visual literacy. The best editors can remix video as fast as you might type.

In fact, the habits of the mashup are borrowed from textual literacy. You cut and paste words on a page. You quote verbatim from an expert. You paraphrase a lovely expression. You add a layer of detail found elsewhere. You borrow the structure from one work to use as your own. You move frames around as if they were phrases. Now you will perform all these literary actions on moving images, in a new visual language.

An image stored on a memory disk instead of celluloid film has a liquidity that allows it to be manipulated as if the picture were words rather than a photo. Hollywood mavericks like George Lucas embraced digital technology early (Lucas founded Pixar) and pioneered a more fluent way of filmmaking. In his *Star Wars* films, Lucas devised a method of moviemaking that has more in common with the way books and paintings are made than with traditional cinematography.

In classic cinematography, a film is planned out in scenes; the scenes are filmed (usually more than once); and from a surfeit of these captured scenes, a movie is assembled. Sometimes a director must go back and shoot "pickup" shots if the final story cannot be told with the available film. With the new screen fluency enabled by digital technology, however, a movie scene is something more malleable—it is like a writer's paragraph, constantly being revised. Scenes are not captured (as in a photo) but built up incrementally, like paint, or text. Layers of visual and audio refinement are added over a crude sketch of the motion, the mix constantly in flux, always changeable. George Lucas's last *Star Wars* movie was layered up in this writerly way. To get the pacing and timing right, Lucas recorded scenes first in crude mock-ups, and then refined them by adding more details and resolution till done. Lightsabers and other effects were digitally painted in, layer by layer. Not a single frame of the final movie was left untouched by manipulation. In essence, his films were written pixel by pixel. Indeed, every single frame in a big-budget Hollywood action film today has been built up with so many layers of additional details that it should be thought of as a moving painting rather than as a moving photograph.

In the great hive mind of image creation, something similar is already happening with still photographs. Every minute, thousands of photographers are uploading their latest photos on websites and apps such as Instagram, Snapchat, WhatsApp, Facebook, and Flickr. The more than 1.5 trillion photos posted so far cover any subject you can imagine; I have not yet been able to stump the sites with an image request that cannot be found. Flickr offers more than half a million images of the Golden Gate Bridge alone. Every conceivable angle, lighting condition, and point of view of the Golden Gate Bridge has been photographed and posted. If you want to use an image of the bridge in your video or movie, there is really no reason to take a new picture of this bridge. It's been done. All you need is a really easy way to find it.

Similar advances have taken place with 3-D models. On the archive for 3-D models generated in the software SketchUp, you can find insanely

detailed three-dimensional virtual models of most major building structures of the world. Need a street in New York? Here's a filmable virtual
set. Need a virtual Golden Gate Bridge? Here it is in fanatical detail, every
rivet in place. With powerful search and specification tools, high-
resolution clips of any bridge in the world can be circulated into the common visual dictionary for reuse. Out of these ready-made "phrases" a film
can be assembled, mashed up from readily available clips or virtual sets.
Media theorist Lev Manovich calls this "database cinema." The databases
of component images form a whole new grammar for moving images.

After all, this is how authors work. We dip into a finite database of
established words, called a dictionary, and reassemble these found words
into articles, novels, and poems that no one has ever seen before. The joy
is recombining them. Indeed, it is a rare author who is forced to invent
new words. Even the greatest writers do their magic primarily by remixing formerly used, commonly shared ones. What we do now with words,
we'll soon do with images.

For directors who speak this new cinematographic language, even
the most photorealistic scenes are tweaked, remade, and written over
frame by frame. Filmmaking is thus liberated from the stranglehold of
photography. Gone is the frustrating method of trying to capture reality
with one or two takes of expensive film and then creating your fantasy
from whatever you get. Here reality, or fantasy, is built up one pixel at a
time as an author would build a novel one word at a time. Photography
exalts the world as it is, whereas this new screen mode, like writing and
painting, is engineered to explore the world as it might be.

But merely producing movies with ease is not enough, just as producing books with ease on Gutenberg's press did not fully unleash text. Real
literacy also required a long list of innovations and techniques that permitted ordinary readers and writers to manipulate text in ways that made
it useful. For instance, quotation symbols make it simple to indicate
where one has borrowed text from another writer. We don't have a

parallel notation in film yet, but we need one. Once you have a large text document, you need a table of contents to find your way through it. That requires page numbers. Somebody invented them in the 13th century. What is the equivalent in video? Longer texts require an alphabetic index, devised by the Greeks and later developed for libraries of books. Someday soon with AI we'll have a way to index the full content of a film. Footnotes, invented in about the 12th century, allow tangential information to be displayed outside the linear argument of the main text. That would be useful in video as well. And bibliographic citations (invented in the 13th century) enable scholars and skeptics to systematically consult sources that influence or clarify the content. Imagine a video with citations. These days, of course, we have hyperlinks, which connect one piece of text to another, and tags, which categorize using a selected word or phrase for later sorting.

All these inventions (and more) permit any literate person to cut and paste ideas, annotate them with her own thoughts, link them to related ideas, search through vast libraries of work, browse subjects quickly, resequence texts, refind material, remix ideas, quote experts, and sample bits of beloved artists. These tools, more than just reading, are the foundations of literacy.

If text literacy meant being able to parse and manipulate texts, then the new media fluency means being able to parse and manipulate moving images with the same ease. But so far, these "reader" tools of visuality have not made their way to the masses. For example, if I wanted to visually compare recent bank failures with similar historical events by referring you to the bank run in the classic movie *It's a Wonderful Life*, there is no easy way to point to that scene with precision. (Which of several sequences did I mean, and which part of them?) I can do what I just did and mention the movie title. I might be able to point to the minute mark for that scene (a new YouTube feature). But I cannot link from this sentence to only those exact "passages" inside an online movie. We don't have the equivalent of a hyperlink for film yet. With true screen fluency, I'd be able to cite specific frames of a film or specific items in a frame. Perhaps I am a historian interested in oriental dress, and I want to refer to a fez worn by someone in the movie *Casablanca*. I should be able to refer to the fez itself (and not the head it is on) by linking to its image as the hat "moves" across many frames, just as I can easily link to a printed reference of the fez in text. Or even better, I'd like to annotate the fez in the film with other film clips of fezzes as references.

With full-blown visuality, I should be able to annotate any object, frame, or scene in a motion picture with any other object, frame, or motion picture clip. I should be able to search the visual index of a film, or peruse a visual table of contents, or scan a visual abstract of its full length. But how do you do all these things? How can we browse a film the way we browse a book?

It took several hundred years for the consumer tools of text literacy to crystallize after the invention of printing, but the first visual literacy tools are already emerging in research labs and on the margins of digital culture. Take, for example, the problem of browsing a feature-length movie. One way to scan a movie would be to super-fast-forward through the two hours in a few minutes. Another way would be to digest it into an abbreviated version in the way a theatrical movie trailer might. Both these methods can compress the time from hours to minutes. But is there a way to reduce the contents of a movie into imagery that could be grasped quickly, as we might see in a table of contents for a book?

Academic research has produced a few interesting prototypes of video summaries, but nothing that works for entire movies. Some popular websites with huge selections of movies (like porn sites) have devised a way for users to scan through the content of full movies quickly in a few seconds. When a user clicks the title frame of a movie, the window skips from one key frame to the next, making a rapid slide show, like a flip book of the movie. The abbreviated slide show visually summarizes a few-hour film in a few seconds. Expert software can be used to identify the key frames in a film in order to maximize the effectiveness of the summary.

The holy grail of visuality is findability—the ability to search the library of all movies the same way Google can search the web, and find a particular focus deep within. You want to be able to type key terms, or simply say, "bicycle plus dog," and then retrieve scenes in any film featuring a dog and a bicycle. In an instant you could locate the moment in *The Wizard of Oz* when the witchy Miss Gulch rides off with Toto. Even better, you want to be able to ask Google to find all the other scenes in all movies similar to that scene. That ability is almost here.

Google's cloud AI is gaining visual intelligence rapidly. Its ability to recognize and remember every object in the billions of personal snapshots that people like me have uploaded is simply uncanny. Give it a picture of a boy riding a motorbike on a dirt road and the AI will label it "boy riding a motorbike on a dirt road." It captioned one photo "two pizzas on a stove," which was exactly what the photo showed. Both Google's and Facebook's AI can look at a photo and tell you the names of the people in it.

Now, what can be done for one image can also be done for moving images, since movies are just a long series of still images in a row. Perceiving movies takes a lot more processing power, in part because there is the added dimension of time (do objects persist as the camera moves?). In a few years we'll be able to routinely search video via AI. As we do, we'll begin to explore the Gutenberg possibilities within moving images. "I consider the pixel data in images and video to be the dark matter of the Internet," says Fei-Fei Li, director of the Stanford Artificial Intelligence Laboratory. "We are now starting to illuminate it."

As moving images become easier to create, easier to store, easier to annotate, and easier to combine into complex narratives, they also become easier to be remanipulated by the audience. This gives images a liquidity similar to words. Fluid images flow rapidly onto new screens, ready to migrate into new media and seep into the old. Like alphabetic bits, they can be squeezed into links or stretched to fit search engines and databases. Flexible images invite the same satisfying participation in both creation and consumption that the world of text does.

In addition to findability, another ongoing revolution within media can be considered "rewindability." In the oral age, when someone spoke, you needed to listen carefully, because once the words were uttered, they were gone. Before the advent of recording technology, there was no backing up, no scrolling back to hear what was missed.

The great historical shift from oral to written communications that occurred thousands of years ago gave the audience (readers) the possibility to scroll back to the beginning of a "speech," by rereading it.

One of the revolutionary qualities of books is their ability to repeat themselves for the reader, at the reader's request, as many times as wanted. In fact, to write a book that is reread is the highest praise for an author. And in many ways authors have exploited this characteristic of books by writing them to be reread. They may add plot points that gain meaning on second reading, hide irony that is only revealed on rereading, or pack it full of details that require close study and rereading to decipher. Vladimir Nabokov once claimed, "One cannot read a book: one can only reread it." Nabokov's novels often featured an unreliable narrator (for instance, *Pale Fire* and *Ada, or Ardor*), which strongly encouraged readers to review the tale from a later, more enlightened perspective. The best mysteries and thrillers tend to end with stealthy last-minute reversals that are brilliantly foreshadowed on second reading. The seven volumes of *Harry Potter* are packed with so many hidden clues that the stories need to be reread for maximum enjoyment.

Our screen-based media in the last century had much in common with books. Movies, like books, are narrative driven and linear. But unlike books, movies were rarely rewatched. Even the most popular blockbusters were released to theaters on a certain day, played in a local theater for a month, and then were rarely seen again, except on late-night television decades later. In the century before videotape, there was no replaying. Television was much the same. A show broadcasted on a schedule. You either watched it at the time or you never saw it. It was uncommon to watch a just released movie more than once, and only a few television episodes would reappear again as summer reruns. And even then, to watch it you needed to schedule your attention to be present on the day and time when that show was due to run.

Because of this "oral" characteristic of movies and television, shows were engineered with the assumption they would be seen only once. That reasonable assumption was made into a feature because it forced the movie's narrative to convey as much as possible in the first impression. But it also diminished it because so much more could be crafted to deliver on second and third encounters.

First the VHS, then DVDs, later TiVos, and now streaming video make it trivially easy to scroll back screenworks. If you want to see something again, you do. Often. If you want to see only a snippet of a movie or television program, you do, at any time. This ability to rewind also applies to commercials, news, documentaries, clips—anything online, in fact. More than anything else, rewindability is what has turned commercials into a new art form. The ability to rewatch them has moved them out of the prison of ephemeral glimpses in the middle of ephemeral shows, to a library of shows that can be read and reread like books. And then shared with others, discussed, analyzed, and studied.

We are now witnessing the same inevitable rewindability of screen-based news. TV news was once an ephemeral stream of stuff that was never meant to be recorded or analyzed—merely inhaled. Now it is rewindable. When we scroll back news, we can compare its veracity, its motives, its assumptions. We can share it, fact-check it, and mix it. Because the crowd can rewind what was said earlier, this changes the posture of politicians, of pundits, of anyone making a claim.

The rewindability of film is what makes 120-hour movies such as *Lost*, or *The Wire*, or *Battlestar Galactica* possible, and enjoyable. They brim with too many details ingeniously molded into them to be apparent on initial viewing; scrolling back at any point is essential.

Music was transformed when it became recorded, rewindable. Live music was meant to be of the moment, and to vary from performance to performance. The ability to scroll back to the beginning and hear music again—that exact performance—changed music forever. Songs became shorter on average, and more melodic and repeatable.

Games now have scroll-back functions that allow replays, redos, or extra lives, a related concept. One can rewind the experience and try again, with slightly different variations, again and again, until one masters that level. On the newest racing games, one can rewind to any previous point by literally running the action backward. All major software packages have an undo button that lets you rewind. The best apps enable unlimited undos, so you can scroll back as far as you want. The most complex pieces of consumer software in existence, such as Photoshop or Illustrator, employ what is called nondestructive editing, which means you can rewind to any particular previous point you want at any time and restart from there, no matter how many changes you've made. The genius of Wikipedia is that it also employs nondestructive editing—all previous versions of an article are kept forever, so any reader can in fact rewind the changes back in time. This "redo" function encourages creativity.

Immersive environments and virtual realities in the future will inevitably be able to scroll back to earlier states. In fact, anything digital will have undo and rewindability as well as remixing.

Going forward, we are likely to get impatient with experiences that don't have undo buttons, such as eating a meal. We can't really replay the taste and smells of a meal. But if we could, that would certainly alter cuisine.

The perfect replication of media in terms of copies is well explored.

But the perfect replication of media in terms of rewinding is less explored. As we begin to lifelog our daily activities, to capture our live streams, more of our lives will be scrollable. Typically I dip into my inbox or outbox several times a day to scroll back to some previous episode of my life. If we expect to scroll back, this will shift what we do the first time. The ability to scroll back easily, precisely, and deeply might change how we live in the future.

In our near future we'll have the option to record as much of our conversations as we care to. It will cost nothing as long as we carry (or wear) a device, and it will be fairly easy to rewind. Some people will record everything as an aid to their memory. The social etiquette around recall will be in flux; private conservations are likely to be off-limits. But more and more of what happens in public will be recorded—and re-viewable—via phone cams, dashboard-mounted webcams on every car, and streetlight-mounted surveillance cams. Police will be required by law to record all activity from their wearables while they are on duty. Rewinding police logs will shift public opinion, just as often vindicating police as not. The everyday routines of politicians and celebrities will be subject to scrolling back from multiple viewpoints, creating a new culture where everyone's past is recallable.

Rewindability and findability are just two Gutenberg-like transformations that moving images are undergoing. These two and many other factors of remixing apply to all newly digitized media, such as virtual reality, music, radio, presentations, and so on.

———

Remixing—the rearrangement and reuse of existing pieces—plays havoc with traditional notions of property and ownership. If a melody is a piece of property you own, like your house, then my right to use it without permission or compensation is very limited. But digital bits are notoriously nontangible and nonrival, as explained earlier. Bits are closer to ideas than to real estate. As far back as 1813, Thomas Jefferson understood that ideas were not really property, or if they were property they differed from real estate. He wrote, "He who receives an idea from me, receives instruction himself without lessening mine; as he who lights his taper at mine, receives light without darkening me." If Jefferson gave you his house at Monticello, you'd have his house and he wouldn't. But if he gave you an idea, you'd have the idea and he'd still have the idea. That weirdness is the source of our uncertainty about intellectual property today.

For the most part our legal system still runs on agrarian principles, where property is real. It has not caught up to the digital era. Not for lack of trying, but because it is difficult to sort out how ownership works in a realm where ownership is less important.

How does one "own" a melody? When you give me a melody, you still have it. Yet in what way is it even yours to begin with if it is one note different from a similar melody a thousand years old? Can one own a note? If you sell me a copy of it, what counts as a copy? What about a backup, or one that streams by? These are not esoteric theoretical questions. Music is one of the major exports of the U.S., a multibillion-dollar industry, and the dilemma of what aspect of intangible music can be owned and how it can be remixed is at the front and center of culture today.

Legal tussles over the right to sample—to remix—snippets of music, particularly when either the sampled song or the borrowing song make a lot of money, are ongoing. The appropriateness of remixing, reusing material from one news source for another is a major restraint for new journalistic media. Legal uncertainty about Google's reuse of snippets from the books it scanned was a major reason it closed down its ambitious book scanning program (although the court belatedly ruled in Google's favor in late 2015). Intellectual property is a slippery realm.

There are many aspects of contemporary intellectual property laws that are out of whack with the reality of how the underlying technology works. For instance, U.S. copyright law gives a temporary monopoly to a creator for his or her creation in order to encourage further creation, but the monopoly has been extended for at least 70 years after the death of the creator, long after a creator's dead body can be motivated by anything. In many cases this unproductive "temporary" monopoly is 100 years long and still being extended longer, and is thus not temporary at all. In a world running at internet speed, a century-long legal lockup is a serious detriment to innovation and creativity. It's a vestigial burden from a former era based on atoms.

The entire global economy is tipping away from the material and toward intangible bits. It is moving away from ownership and toward access. It is tilting away from the value of copies and toward the value of networks. It is headed for the inevitability of constant, relentless, and increasing remixing. The laws will be slow to follow, but they will follow.

So what should the new laws favor in a world of remixing?

Appropriation of existing material is a venerable and necessary practice. As the economists Romer and Arthur remind us, recombination is really the only source of innovation—and wealth. I suggest we follow the question, "Has it been transformed by the borrower?" Did the remixing, the mashup, the sampling, the appropriation, the borrowing—did it transform the original rather than just copy it? Did Andy Warhol transform the Campbell's soup can? If yes, then the derivative is not really a "copy"; it's been transformed, mutated, improved, evolved. The answer each time is still a judgment call, but the question of whether it has been transformed is the right question.

Transformation is a powerful test because "transformation" is another term for becoming. "Transformation" acknowledges that the creations we make today will become, and should become, something else tomorrow. Nothing can remain untouched, unaltered. By that I mean, every creation that has any value will eventually and inevitably be transformed—in some version—into something different. Sure, the version of *Harry*

Potter that J. K. Rowling published in 1997 will always be available, but it is inevitable that another thousand fan fiction versions of her book will be penned by avid amateurs in the coming decades. The more powerful the invention or creation, the more likely and more important it is that it will be transformed by others.

In 30 years the most important cultural works and the most powerful mediums will be those that have been remixed the most.

9

INTERACTING

Virtual reality (VR) is a fake world that feels absolutely authentic. You can experience a hint of VR when you watch a movie in 3-D on a jumbo IMAX screen in surround sound. At moments you'll be fully immersed in a different world, which is what virtual reality aims for. But this movie experience is not full VR, because while your imagination travels to another place in a theater, your body doesn't. It *feels* like you are in a chair. Indeed, in a theater you must remain sitting in the same spot looking straight ahead passively in order for the immersive magic to work.

A much more advanced VR experience might be like the world Neo confronts in the movie *The Matrix*. Even as Neo runs, leaps, and battles a hundred clones in a computerized world, it feels totally real to him. Maybe even hyperreal—realer than real. His vision, hearing, and touch are hijacked by the synthetic world so completely that he cannot detect its artificiality. A yet even more advanced mode of VR is the holodeck on *Star Trek*. There, holographic projections of objects and people are so real in fiction they are solid to the touch. A simulated environment that you can enter at will is a recurring science fiction dream that is long overdue.

Today's virtual reality is in between the elemental feeling of a 3-D IMAX movie and the ultimate holodeck simulation. A VR experience in 2016 can involve a billionaire's mansion in Malibu that you walk through, room by overstuffed room, feeling as if you are really there when you are actually standing a thousand miles away wearing a helmet in a real estate agent's office. That is something I experienced recently. Or it might be a fantasy world full of prancing unicorns where you authentically feel you are flying, once you put on special glasses. Or it may be an alternate version of the office cubicle you are sitting in that includes floating touch screens and an avatar of a distant coworker speaking next to you. In each case, you have a very strong sense that you are physically present in this virtual world, in large part because you can do things—look around, freely move in any direction, move objects—that persuade you that you are "really there."

Recently I've had the opportunity to immerse myself in many prototype VR worlds. The best of these achieve an unshakeable sense of presence. The usual goal for increasing the degree of realism while you tell a story is to suspend disbelief. The goal for VR is not to suspend belief but to ratchet up belief—that you are somewhere else, and maybe even somebody else. Even if your intellectual mind can figure out you are really in a swivel chair, your embodied "I" will be convinced you are trudging through a swamp.

For the past decade, researchers inventing VR have settled on a standard demonstration of this overpowering presence. The visitor waiting for the demo stands in the center of an actual real nondescript waiting room. A pair of large dark goggles rest on a stool. The visitor dons the goggles and is immediately immersed into a virtual version of the same room she was standing in, with the same nondescript paneling and chairs. Not much is changed from her point of view. She can look around. The scene looks a little coarser through the goggles. But slowly the floor of the room begins to drop away, leaving the visitor standing on a plank that now floats over the receding floor 30 meters below. She is asked to walk out farther on the plank suspended high over a most realistic pit. The realism of the scene has been improved over the years so that by now the response of the visitor is very predictable. Either she cannot move her feet or she trembles as she inches forward, palms sweating.

When I was plunged into this scene myself, I reacted the same way. My mind reeled. My conscious mind kept whispering to me that I was in a dim room in the research labs of Stanford, but my primitive mind had hijacked my body. It was insisting that I was perched on a too narrow plank too high in the sky and that I must back off this plank immediately. Right now! My fear of heights kicked in. My knees began to shake. I was almost nauseous. Then I did something stupid. I decided to jump off the plank a little ways down onto a nearby ledge in the virtual world. But of course there was no "down," so my real body dove onto the floor. But since I was actually standing on the floor, I was caught as I fell by two strong spotters in the real room, who were standing there precisely for this purpose. My reaction was completely normal; almost everyone falls.

Totally believable virtual reality is just about here. But I have been wrong about VR before. In 1989 a friend of a friend invited me to his lab in Redwood City, California, to see some gear he had invented. The lab turned out to be a couple of rooms in an office complex that were missing most of their desks. The walls were covered by a gallery of neoprene body-suits embroidered with wires, large gloves sporting electronic compo-nents, and rows of duct-taped swimming goggles. The guy I'd gone to see, Jaron Lanier, sported shoulder-length blond dreadlocks. I wasn't sure where this was going, but Jaron promised me a new experience, some-thing he called virtual reality.

A few minutes later Lanier handed me one black glove, a dozen wires snaking from the fingers across the room to a standard desktop PC. I put it on. Lanier then placed a set of black goggles suspended by a web of straps onto my head. A thick black cable ran down my back from the headgear to his computer. Once my eyes focused inside the goggles, I was in. I was inside a place bathed in a diffuse light blue. I could see a cartoon version of my glove in the exact place my real hand felt it was. The virtual glove moved in sync with my hand. It was now "my" glove, and I felt—in my body, not just my head—very strongly that I was not in an office. Lanier himself then climbed into his own creation. Using his own helmet and glove, he appeared in his own world as a girl avatar, since the beauty of his system was that you could design your avatar to look like anything you wanted. Two of us now inhabited this first mutual dream space. In 1989.

Lanier popularized the term "virtual reality," but he was not the only person working on immersive simulations at that time in the late 1980s. Several universities, a few startups, as well as the U.S. military had comparable prototypes, some with slightly different approaches for creating the phenomenon. I felt I had seen the future during my plunge into his microcosmos and wanted as many of my friends and fellow pundits as possible to experience what I had. With the help of the magazine I was then editing (*Whole Earth Review*), we organized the first public demo of every VR rig that existed in the fall of 1990. For 24 hours, from Saturday noon to Sunday noon, anyone who bought a ticket could stand in line to try out as many of the two dozen or so VR prototypes as they could. In the wee hours of the night I saw the psychedelic champion Tim Leary compare VR to LSD. The overwhelming impression spun by the buggy gear was total plausibility. These simulations were real. The views were coarse, the vision often stuttered, but the intended effect was inarguable: You went somewhere else. The next morning William Gibson, an up-and-coming science fiction writer who stayed up the night testing cyberspace for the first time, was asked what he thought about these new portals to synthetic worlds. He then first uttered his now famous remark: "The future is already here; it's just not evenly distributed."

VR was so uneven, however, it faded. The next steps never happened.

All of us, myself included, thought VR technology would be ubiquitous in five years or so—at least by the year 2000. But no advances happened till 2015, 25 years after Jaron Lanier's pioneering work. The particular problem with VR was that close enough was not close enough. For extended stays in VR longer than 10 minutes, the coarseness and stuttering motion caused nausea. The cost of gear sufficiently powerful, fast, and comfortable enough to overcome nausea was many tens of thousands of dollars. Therefore VR remained out of reach to consumers, and also out of reach for many startup developers who needed to jump-start the creation of VR content to spark the purchase of the gear.

Twenty-five years later a most unlikely savior appeared: phones! The runaway global success of the smartphone drove the quality of their tiny hi-res screens way up and their cost way down. The eye screens for a VR goggle are approximately the size and resolution of a smartphone screen, so today VR headsets are basically built out of cheap phone screen technology. At the same time, motion sensors in phones followed the same path of increasing performance and decreasing cost, until these motion sensors could be borrowed by VR displays to track head, hand, and body positions for very little. In fact, the first consumer VR models from Samsung and Google use a regular smartphone slipped into an empty head-mounted display unit. Put a Samsung Gear VR on and you look into a phone; your movements are tracked by the phone, so the phone sends you into an alternative world.

It's not difficult to see how VR will soon triumph in movies of the future, particularly visceral genres like horror, erotic, or thrillers—where your gut is also caught up in the story. It's also easy to imagine VR occupying a prime role in video games. No doubt hundreds of millions of avid players will eagerly don a suit, gloves, and helmet and then teleport to a far-away place to hide, shoot, kill, and explore, either solo or in tight bands of friends. Of course, the major funder of consumer VR development today is the game industry. But VR is much bigger than this.

———

Two benefits propel VR's current rapid progress: presence and interaction. "Presence" is what sells VR. All the historical trends in cinema technology bend toward increased realism, starting from sound, to color, to 3-D, to faster, smoother frame rates. Those trends are now being accelerated inside VR. Week by week the resolution increases, the frame rate jumps, the contrast deepens, the color space widens, and the high-fidelity sound sharpens, all of it improving faster than it does on big screens. That is, VR is getting more "realistic" faster than movies are. Within a decade, when you look into a state-of-the-art virtual reality display, your eye will be fooled into thinking you are looking through a real window into a real world. It'll be bright—no flicker, no visible pixels. You will feel this is absolutely for sure real. Except it isn't.

The second generation of VR technology relies on a new, innovative "light field" projection. (The first commercial light field units are the HoloLens made by Microsoft and Magic Leap funded by Google.) In this design the VR is projected onto a semi-transparent visor much like a holograph. This permits the projected "reality" to overlay the reality you see normally without goggles. You could be standing in your kitchen and see the robot R2-D2 right before you in perfect resolution. You could walk around it, get closer, even move it to inspect it, and it would retain its authenticity. This overlay is called augmented reality (AR). Because the artificial part is added to your ordinary view of the world, your eyes are focused deeper than they are on a screen near your eyes, so this technological illusion is packed with presence. You almost swear it is really there.

Microsoft's vision for light field AR is to build the office of the future. Instead of workers sitting in a cubicle in front of a wall of monitor screens, they sit in an open office wearing HoloLenses and see a huge wall of virtual screens around them. Or they click to be teleported to a 3-D conference room with a dozen coworkers who live in different cities. Or they click to a training room where an instructor will walk them though a first-aid class, guiding their avatars through the proper procedures. "See this? Now you do it." In most ways, the AR class will be superior to a real-world class.

The reason why cinematic realism is advancing faster in VR than in cinema itself is due to a neat trick performed by head-mounted displays. To fill a gigantic IMAX cinema screen with the proper resolution and brightness to convince you it is a mere window into reality requires a massive amount of computation and luminosity. To fill a 60-inch flat screen with the same window-clear realism is a smaller challenge, but still daunting. It is much easier to get a tiny visor in front of your face up to that quality. Because a head-mounted display follows your gaze no matter where you look—it is always in front of your eyes—you see full realism all the time. Therefore if you make fully 3-D clear-as-a-window vision and keep it in view no matter where you look, you can create a virtual IMAX inside of the VR. Turn your gaze anywhere on the screen and the realism follows your gaze because the tech is physically attached to your face. In fact, the entire 360-degree virtual world appears in the same ultimate resolution as what's in front of your eyes. And since what is in front of your eyes is just a small surface area, it is much easier and cheaper to magnify small improvements in quality. This tiny little area can invoke a huge disruptive presence.

But while "presence" will sell it, VR's enduring benefits spring from its interactivity. It is unclear how comfortable, or uncomfortable, we'll be with the encumbrances of VR gear. Even the streamlined Google Glass (which I also tried), a very mild AR display not much bigger than sunglasses, seemed too much trouble for most people in its first version. Presence will draw users in, but it is the interactivity quotient of VR that will keep it going. Interacting in all degrees will spread out to the rest of the technological world.

About 10 years ago, Second Life was a fashionable destination on the internet. Members of Second Life created full-body avatars in a simulated world that mirrored "first life." A lot of their time was spent remaking their avatars into beautiful people with glamorous clothes and socializing with other members' incredibly beautiful avatars. Members devoted lifetimes to building super beautiful homes and slick bars and discos. The environment and avatars were created in full 3-D, but due to technological constraints, members could only view the world in flat 2-D on their desktop screens. (Second Life is rebooting itself as a 3-D world in 2016, code-named Project Sansa.) Avatars communicated via text balloons floating over their heads, typed by owners. It was like walking around in a comic book. This clunky interface held back any deep sense of presence. The main attraction of Second Life was the completely open space for constructing a quasi-3-D environment. Your avatar walked onto an empty plain, like the blank field at a Burning Man festival, and could begin constructing the coolest and most outrageous buildings, rooms, or wilderness places. Physics didn't matter, materials were free, anything was possible. But it took many hours to master the arcane 3-D tools. In 2009 a game company in Sweden, Minecraft, launched a similar construction world in quasi-3-D, but employed idiot-easy building blocks stacked like giant Legos. No learning was necessary. Many would-be builders migrated to Minecraft.

Second Life's success had risen on the ability of kindred creative spirits to socialize, but when the social mojo moved to the mobile world, no phones had enough computing power to handle Second Life's sophisticated 3-D, so the biggest audiences moved on. Even more headed to Minecraft, whose crude low-res pixelation allowed it to run on phones. Millions of members are still loyal to Second Life, and today at any hour about 50,000 avatars are simultaneously roaming the imaginary 3-D worlds built by users. Half of them are there for virtual sex, which relies more on the social component than on realism. A few years ago the founder of Second Life, Phil Rosedale, started another VR-ish company trying to harness the social opportunities of an open simulated world and to invent a more convincing VR.

Recently I visited the offices of Rosedale's startup, High Fidelity. As the name implies, the aim of its project is to raise the realism in virtual worlds occupied by thousands—maybe tens of thousands—of avatars at once. Create a realistic thriving virtual city. Jaron Lanier's pioneering VR permitted two occupants at once, and the thing I noticed (and everyone else who visited) was that other people in VR were far more interesting than other things. Experimenting again in 2015, I found the best demos of synthetic worlds are ones that trigger a deep presence not with the most pixels per inch, but with the most engagement of other people. To that end, High Fidelity is exploiting a neat trick. Taking advantage of the tracking abilities of cheap sensors, it can mirror the direction of your gaze in both worlds. Not just where you turn your head, but where you turn your eyes. Nano-small cameras buried *inside* the headset look back at your real eyes and transfer your exact gaze onto your avatar. That means that if someone is talking to your avatar, their eyes are staring at your eyes, and yours at theirs. Even if you move, requiring them to rotate their head, their eyes continue to lock onto yours. This eye contact is immensely magnetic. It stirs intimacy and radiates a felt presence.

Nicholas Negroponte, head of MIT's Media Lab, once quipped in the 1990s that the urinal in the men's restroom was smarter than his computer because it knew he was there and would flush when he left, while his computer had no idea he was sitting in front of it all day. That is still kind of true today. Laptops and even tablets and phones are largely ignorant of their owners' use of them. That is starting to change with cheap eye tracking mechanisms like the one in the VR headsets. The newest Samsung Galaxy phone contains eye tracking technology so the phone knows precisely where on the screen you are looking. Gaze tracking can be used in many ways. It can speed up screen navigation since you often look at something before your finger or mouse moves to confirm it. Also, by measuring the duration of thousands of people's gazes on a screen, software can generate maps that rank areas of greater or lesser attention. Website owners can then discern what part of their front page people actually look at and what parts are glanced over, and use that information to improve the design. An app maker can use gaze patterns of visitors to find which parts of an app's interface demand too much attention, suggesting a difficulty that needs to be fixed. Mounted in a dashboard in a car, the same gaze technology can detect when drivers are drowsy or distracted.

The tiny camera eyes that now stare back at us from any screen can be trained with additional skills. First the eyes were trained to detect a generic face, used in digital cameras to assist focusing. Then they were taught to detect particular faces—say, yours—as identity passwords. Your laptop looks into your face, and deeper into your irises, to be sure it is you before it opens its home page. Recently researchers at MIT have taught the eyes in our machines to detect human emotions. As we watch the screen, the screen is watching us, where we look, and how we react. Rosalind Picard and Rana el Kaliouby at the MIT Media Lab have developed software so attuned to subtle human emotions that they claim it can detect if someone is depressed. It can discern about two dozen different emotions. I had a chance to try a beta version of this "affective technology," as Picard calls it, on Picard's own laptop. The tiny eye in the lid of her laptop peering at me could correctly determine if I was perplexed or engaged with a difficult text. It could tell if I was distracted while viewing a long video. Since this perception is in real time, the smart software can adapt it to what I'm viewing. Say I am reading a book and my frown shows I've stumbled on a certain word; the text could expand a definition. Or if it realizes I am rereading the same passage, it could supply an annotation for that passage. Similarly, if it knows I am bored by a scene in a video, it could jump ahead or speed up the action.

We are equipping our devices with senses—eyes, ears, motion—so that we can interact with them. They will not only know we are there, they will know who is there and whether that person is in a good mood. Of course, marketers would love to get hold of our quantified emotions, but this knowledge will serve us directly as well, enabling our devices to respond to us "with sensitivity" as we hope a good friend might.

In the 1990s I had a conversation with the composer Brian Eno about the rapid changes in music technology, particularly its sprint from analog to digital. Eno made his reputation by inventing what we might now call electronic music, so it was a surprise to hear him dismiss a lot of digital instruments. His primary disappointment was with the instruments' atrophied interfaces—little knobs, sliders, or tiny buttons mounted on square black boxes. He had to interact with them by moving only his fingers. By comparison, the sensual strings, table-size keyboards, or meaty drumheads of traditional analog instruments offered more nuanced bodily interactions with the music. Eno told me, "The trouble with computers is that there is not enough Africa in them." By that he meant that interacting with computers using only buttons was like dancing with only your fingertips, instead of your full body, as you would in Africa.

Embedded microphones, cameras, and accelerometers inject some Africa into devices. They provide embodiment in order to hear us, see us, feel us. Swoosh your hand to scroll. Wave your arms with a Wii. Shake or tilt a tablet. Let us embrace our feet, arms, torso, head, as well as our fingertips. Is there a way to use our whole bodies to overthrow the tyranny of the keyboard?

One answer first premiered in the 2002 movie *Minority Report*. The director, Steven Spielberg, was eager to convey a plausible scenario for the year 2050, and so he convened a group of technologists and futurists to brainstorm the features of everyday life in 50 years. I was part of that invited group, and our job was to describe a future bedroom, or what music would sound like, and especially how you would work on a computer in 2050. There was general consensus that we'd use our whole bodies and all our senses to communicate with our machines. We'd add Africa by standing instead of sitting. We think different on our feet. Maybe we'd add some Italy by talking to machines with our hands. One of our group, John Underkoffler, from the MIT Media Lab, was way ahead in this scenario and was developing a working prototype using hand motions to control data visualizations. Underkoffler's system was woven into the film. The Tom Cruise character stands, raises his hands outfitted with a VR-like glove, and shuffles blocks of police surveillance data, as if conducting music. He mutters voice instructions as he dances with the data. Six years later, the *Iron Man* movies picked up this theme. Tony Stark, the protagonist, also uses his arms to wield virtual 3-D displays of data projected by computers, catching them like a beach ball, rotating bundles of information as if they were objects.

It's very cinematic, but real interfaces in the future are far more likely to use hands closer to the body. Holding your arms out in front of you for more than a minute is an aerobic exercise. For extended use, interaction will more closely resemble sign language. A future office worker is not going to be pecking at a keyboard—not even a fancy glowing holographic keyboard—but will be talking to a device with a newly evolved set of hand gestures, similar to the ones we now have of pinching our fingers in to reduce size, pinching them out to enlarge, or holding up two L-shaped pointing hands to frame and select something. Phones are very close to perfecting speech recognition today (including being able to translate in real time), so voice will be a huge part of interacting with devices. If you'd like to have a vivid picture of someone interacting with a portable device in the year 2050, imagine them using their eyes to visually "select" from a set of rapidly flickering options on the screen, confirming with lazy audible grunts, and speedily fluttering their hands in their laps or at their waist. A person mumbling to herself while her hands dance in front of her will be the signal in the future that she is working on her computer.

Not only computers. All devices need to interact. If a thing does not interact, it will be considered broken. Over the past few years I've been collecting stories of what it is like to grow up in the digital age. As an example, one of my friends had a young daughter under five years old. Like many other families these days, they didn't have a TV, just computing screens. On a visit to another family who happened to have a TV, his daughter gravitated to the large screen. She went up to the TV, hunted around below it, and then looked behind it. "Where's the mouse?" she asked. There had to be a way to interact with it. Another acquaintance's son had access to a computer starting at the age of two. Once, when she and her son were shopping in a grocery store, she paused to decipher the label on a product. "Just click on it," her son suggested. Of course cereal boxes should be interactive! Another young friend worked at a theme park. Once, a little girl took her picture, and after she did, she told the park worker, "But it's not a real camera—it doesn't have the picture on the back." Another friend had a barely speaking toddler take over his iPad. She could paint and easily handle complicated tasks on apps almost before she could walk. One day her dad printed out a high-resolution image on photo paper and left it on the coffee table. He noticed his toddler came up and tried to unpinch the photo to make it larger. She tried unpinching it a few times, without success, and looked at him, perplexed. "Daddy, broken." Yes, if something is not interactive, it is broken.

The dumbest objects we can imagine today can be vastly improved by outfitting them with sensors and making them interactive. We had an old standard thermostat running the furnace in our home. During a remodel we upgraded to a Nest smart thermostat, designed by a team of ex-Apple execs and recently bought by Google. The Nest is aware of our presence. It senses when we are home, awake or asleep, or on vacation. Its brain, connected to the cloud, anticipates our routines, and over time builds up a pattern of our lives so it can warm up the house (or cool it down) just a few minutes before we arrive home from work, turn it down after we leave, except on vacations or on weekends, when it adapts to our schedule. If it senses we are unexpectedly home, it adjusts itself. All this watching of us and interaction optimizes our fuel bill.

One consequence of increased interaction between us and our artifacts is a celebration of an artifact's embodiment. The more interactive it is, the more it should sound and feel beautiful. Since we might spend hours holding it, craftsmanship matters. Apple was the first to recognize that this appetite applies to interactive goods. The gold trim on the Apple Watch is to feel. We end up caressing an iPad, stroking its magic surface, gazing into it for hours, days, weeks. The satin touch of a device's surface, the liquidity of its flickers, the presence or lack of its warmth, the quality of its build, the temperature of its glow will come to mean a great deal to us.

What could be more intimate and interactive than wearing something that responds to us? Computers have been on a steady march toward us. At first computers were housed in distant air-conditioned basements, then they moved to nearby small rooms, then they crept closer to us perched on our desks, then they hopped onto our laps, and recently they snuck into our pockets. The next obvious step for computers is to lay against our skin. We call those wearables.

We can wear special spectacles that reveal an augmented reality. Wearing such a transparent computer (an early prototype was Google Glass) empowers us to see the invisible bits that overlay the physical world. We can inspect a cereal box in the grocery store and, as the young boy suggested, simply click it within our wearable to read its meta-information. Apple's watch is a wearable computer, part health monitor, but mostly a handy portal to the cloud. The entire super-mega-processing power of the entire internet and World Wide Web is funneled through that little square on your wrist. But wearables in particular mean smart clothes. Of course, itsy-bitsy chips can be woven into a shirt so that the shirt can alert a smart washing machine to its preferred washing cycles, but wearables are more about the wearer. Experimental smart fabrics such as those from Project Jacquard (funded by Google) have conductive threads and thin flexible sensors woven into them. They will be sewn into a shirt you interact with. You use fingers of one hand to swipe the sleeve of your other arm the way you'd swipe an iPad, and for the same reason: to bring up something on a screen or in your spectacles. A smart shirt like the Squid, a prototype from Northeastern University, can feel—in fact measure—your posture, recording it in a quantified way, and then actuating "muscles" in the shirt that contract precisely to hold you in the proper posture, much as a coach would. David Eagleman, a neuroscientist at Baylor College, in Texas, invented a supersmart wearable vest that translates one sense into another. The Sensory Substitution Vest takes audio from tiny microphones in the vest and translates those sound waves into a grid of vibrations that can be felt by a deaf person wearing it. Over a matter of months, the deaf person's brain reconfigures itself to "hear" the vest vibrations as sound, so by wearing this interacting cloth, the deaf can hear.

You may have seen this coming, but the only way to get closer than wearables over our skin is to go under our skin. Jack into our heads. Directly connect the computer to the brain. Surgical brain implants really do work for the blind, the deaf, and the paralyzed, enabling the handicapped to interact with technology using only their minds. One experimental brain jack allowed a quadriplegic woman to use her mind to control a robotic arm to pick up a coffee bottle and bring it to her lips so she could drink from it. But these severely invasive procedures have not been tried to enhance a healthy person yet. Brain controllers that are noninvasive have already been built for ordinary work and play, and they do work. I tried several lightweight brain-machine interfaces (BMIs) and I was able to control a personal computer simply by thinking about it. The apparatus generally consists of a hat of sensors, akin to a minimal bicycle helmet, with a long cable to the PC. You place it on your head and its many sensor pads sit on your scalp. The pads pick up brain waves, and with some biofeedback training you can generate signals at will. These signals can be programmed to perform operations such as "Open program," "Move mouse," and "Select this." You can learn to "type." It's still crude, but the technology is improving every year.

In the coming decades we'll keep expanding what we interact with. The expansion follows three thrusts.

1. More senses

We will keep adding new sensors and senses to the things we make. Of course, everything will get eyes (vision is almost free), and hearing, but one by one we can add superhuman senses such as GPS location sensing, heat detection, X-ray vision, diverse molecule sensitivity, or smell. These permit our creations to respond to us, to interact with us, and to adapt themselves to our uses. Interactivity, by definition, is two way, so this sensing elevates our interactions with technology.

2. More intimacy

The zone of interaction will continue to march closer to us. Technology will get closer to us than a watch and pocket phone. Interacting will be more intimate. It will always be on, everywhere. Intimate technology is a wide-open frontier. We think technology has saturated our private space, but we will look back in 20 years and realize it was still far away in 2016.

3. More immersion

Maximum interaction demands that we leap into the technology itself. That's what VR allows us to do. Computation so close that we are inside it. From within a technologically created world, we interact with each other in new ways (virtual reality) or interact with the physical world in a new way (augmented reality). Technology becomes a second skin.

Recently I joined some drone hobbyists who meet in a nearby park on Sundays to race their small quadcopters. With flags and foam arches they map out a course over the grass for their drones to race around. The only way to fly drones at this speed is to get inside them. The hobbyists mount tiny eyes at the front of their drones and wear VR goggles to peer through them for what is called a first-person view (FPV). They are now the drone. As a visitor I don an extra set of goggles that piggyback on their camera signals and so I find myself sitting in the same pilots' seats and see what each pilot sees. The drones dart in, out, and around the course obstacles, chasing each other's tails, bumping into other drones, in scenes reminiscent of a *Star Wars* pod race. One young guy who's been flying radio control model airplanes since he was a boy said that being able to immerse himself into the drone and fly from inside was the most sensual experience of his life. He said there was almost nothing more pleasurable than actually, really free flying. There was no virtuality. The flying experience was real.

————

The convergence of maximum interaction plus maximum presence is found these days in free-range video games. For the past several years I've been watching my teenage son play console video games. I am not twitchy enough myself to survive more than four minutes in a game's alterworld, but I find I can spend an hour just watching the big screen as my son encounters dangers, shoots at bad guys, or explores unknown territories and dark buildings. Like a lot of kids his age, he's played the classic shooter games like *Call of Duty*, *Halo*, and *Uncharted 2*, which have scripted scenes of engagement. However, my favorite game as a voyeur is the now dated game *Red Dead Redemption*. This is set in the vast empty country of the cowboy West. Its virtual world is so huge that players spend a lot of time on their horses exploring the canyons and settlements, searching for clues, and wandering the land on vague errands. I'm happy to ride alongside as we pass through frontier towns in pursuit of his quests. It's a movie you can roam in. The game's open-ended architecture is similar to the very popular *Grand Theft Auto*, but it's a lot less violent. Neither of us knows what will happen or how things will play out.

There are no prohibitions about where you can go in this virtual place. Want to ride to the river? Fine. Want to chase a train down the tracks? Fine. How about ride up alongside the train and then hop on and ride inside the train? OK! Or bushwhack across sagebrush wilderness from one town to the next? You can ride away from a woman yelling for help or—your choice—stop to help her. Each act has consequences. She may need help or she may be bait for a bandit. One reviewer speaking of the interacting free will in the game said: "I'm sincerely and pleasantly surprised that I can shoot my own horse in the back of the head while I'm riding him, and even skin him afterward." The freedom to move in any direction in a seamless virtual world rendered with the same degree of fidelity as a Hollywood blockbuster is intoxicating.

It's all interactive details. Dawns in the territory of *Red Dead Redemption* are glorious, as the horizon glows and heats up. Weather forces itself on the land, which you sense. The sandy yellow soil darkens with appropriate wet splotches as the rain blows down in bursts. Mist sometimes drifts in to cover a town with realistic veiling, obscuring shadowy figures. The pink tint of each mesa fades with the clock. Textures pile up. The scorched wood, the dry brush, the shaggy bark—every pebble or twig—is rendered in exquisite minutiae at all scales, casting perfect overlapping shadows that make a little painting. These nonessential finishes are surprisingly satisfying. The wholesale extravagance is compelling.

The game lives in a big world. A typical player might take around 15 or so hours to zoom through once, while a power player intent on achieving all the game rewards would need 40 to 50 hours to complete it. At every step you can choose any direction to take the next step, and the next, and next, and yet the grass under your feet is perfectly formed and every blade detailed, as if its authors anticipated you would tread on this microscopic bit of the map. At any of a billion spots you can inspect the details closely and be rewarded, but most of this beauty will never be seen. This warm bath of freely given abundance triggers a strong conviction that this is "natural," that this world has always been, and that it is good. The overall feeling inside one of these immaculately detailed, stunningly interactive worlds stretching to the horizons is of being immersed in completeness. Your logic knows this can't be true, but as on the plank over the pit, the rest of you believes it. This realism is just waiting for the full immersion of VR interaction. At the moment, the spatial richness of these game worlds must be viewed in 2-D.

Cheap, abundant VR will be an experience factory. We'll use it to visit environments too dangerous to risk in the flesh, such as war zones, deep seas, or volcanoes. Or we'll use it for experiences we can't easily get to as humans—to visit the inside of a stomach, the surface of a comet. Or to swap genders, or become a lobster. Or to cheaply experience something expensive, like a flyby of the Himalayas. But experiences are generally not sustainable. We enjoy travel experiences in part because we are only visiting briefly. VR, at least in the beginning, is likely to be an experience we dip in and out of. Its presence is so strong we may want it only in small, measured doses. But we have no limit on the kind of interacting we crave.

These massive video games are pioneering new ways of interacting. The total interactive freedom suggested by unlimited horizons is illusionary in these kinds of games. Players, or the audience, are assigned tasks to accomplish and given motivations to stay till the end. Actions in the game are channeled funnel-like to meet the next bottleneck of the overall narrative, so the game eventually reveals a destiny, but your choices as a player still matter in what kind of points you accumulate. There's a tilt in the overall world, so no matter how many explorations you make, you tend to drift over time toward an inevitable incident. When the balance between an ordained narrative and freewill interaction is tweaked just right, it creates the perception of great "game play"—a sweet feeling of being part of something large that is moving forward (the game's narrative) while you still get to steer (the game's play).

The games' designers tweak the balance, but the invisible force that nudges players in certain directions is an artificial intelligence. Most of the action in open-ended games like *Red Dead Redemption*, especially the interactions of supporting characters, is already animated by AI. When you halt at a random homestead and chat with the cowhand, his responses are plausible because in his heart beats an AI. AI is seeping into VR and AR in other ways as well. It will be used to "see" and map the physical world you are really standing in so that it can transport you to a synthetic world. That includes mapping your physical body's motion. An AI can watch you as you sit, stand, move around in, say, your office without the need of special tracking equipment, then mirror that in the virtual world. An AI can read your route through the synthetic environment and calculate interferences needed to herd you in certain directions, as a minor god might do.

Implicit in VR is the fact that everything—without exception—that occurs in VR is tracked. The virtual world is defined as a world under total surveillance, since nothing happens in VR without tracking it first. That makes it easy to gameify behavior—awarding points, or upping levels, or scoring powers, etc.—to keep it fun. However, today the physical world is so decked out with sensors and interfaces that it has become a parallel tracking world. Think of our sensor-filled real world as a nonvirtual virtual reality that we spend most of our day in. As we are tracked by our surroundings and indeed as we track our quantified selves, we can use the same interaction techniques that we use in VR. We'll communicate with our appliances and vehicles using the same VR gestures. We can use the same gameifications to create incentives, to nudge participants in preferred directions in real life. You might go through your day racking up points for brushing your teeth properly, walking 10,000 steps, or driving safely, since these will all be tracked. Instead of getting A-pluses on daily quizzes, you level up. You get points for picking up litter or recycling. Ordinary life, not just virtual worlds, can be gameified.

The first technological platform to disrupt a society within the lifespan of a human individual was personal computers. Mobile phones were the second platform, and they revolutionized everything in only a few decades. The next disrupting platform—now arriving—is VR. Here is how a day plugged into virtual and augmented realities may unfold in the very near future.

I am in VR, but I don't need a headset. The surprising thing that few people expected way back in 2016 is that you don't need to wear goggles, or even a pair of glasses, in order to get a basic "good enough" augmented reality. A 3-D image projects directly into my eyes from tiny light sources that peek from the corner of my rooms, all without the need of something in front of my face. The quality is good enough for most applications, of which there are tens of thousands.

The very first app I got was the ID overlay. It recognizes people's faces and then displays their name, association, and connection to me, if any. Now that I am used to this, I can't roam outside without it. My friends say some quasi-legal ID apps provide a lot more immediate information about strangers, but you need to be wearing gear that keeps what you see private—otherwise you'll get tagged for rude behavior.

I wear a pair of AR glasses outside to get a sort of X-ray view of my world. I use it first to find good connectivity. The warmer the colors in the world, the closer I am to heavy-duty bandwidth. With AR on I can summon earlier historical views layered on top of whatever place I am looking at, a nifty trick I used extensively in Rome. There, a fully 3-D life-size intact Colosseum appeared synchronized over the ruins as I clambered through them. It's an unforgettable experience. It also shows me comments virtually "nailed" to different spots in the city left by other visitors that are viewable only from that very place. I left a few notes in spots for others to discover as well. The app reveals all the underground service pipes and cables beneath the street, which I find nerdly fascinating. One of the weirder apps I found is one that will float the dollar value—in big red numbers—over everything you look at. Almost any subject I care about has an overlay app that displays it as an apparition. A fair amount of public art is now 3-D mirages. The plaza in our town square hosts an elaborate rotating 3-D projection that is refreshed twice a year, like a museum art show. Most of the buildings downtown are reskinned with alternative facades inside AR, each facade commissioned by an architect or artist. The city looks different each time I walk through it.

I wore VR goggles all through high school. These lightweight frames give a much more vivid image than glassless AR. In class I'd watch all kinds of simulations, especially how-to rehearsals. I preferred the "ghost" mode in maker classes, like cooking or electrical hacking. That is how I learned how to weld. In AR I slipped my hands into the position of the teacher's ghostly virtual guide hands in order to correctly grip the virtual welding rod held against the virtual steel tube. I tried to move my hands to follow the ghost hands. My virtual welds were only as good as my actions. For sports I wore a full helmet display. I rehearsed my moves with 360-degree motion on a real field, shadowing a model shadow body. I also spend a lot of time practicing plays in VR in a room. A couple of sports, like broadswording, we played entirely inside VR.

At my "office" I wear an AR visor on my forehead. The visor is a curved band about hand width wide that is held a few inches away from my eyes for extra comfort during daylong use. The powerful visor throws up virtual screens all around me. I have about 12 virtual screens of all sizes and large data sets I can wrestle with my hands. The visor provides enough resolution and speed that most of my day I am communicating with virtual colleagues. But I see them in a real room, so I am fully present in reality as well. Their photorealistic 3-D avatar captures their life-size likeness accurately. My coworkers and I usually sit at a virtual table in a real room while we work independently, but we can walk around each other's avatar. We converse and overhear each other just as if we are in the same room. It is so convenient to pop up an avatar that even if my real coworker is on the other side of the real room, we'll just meet in the AR rather than walk across the gap.

When I want to get really serious about augmented reality, I'll wear an AR roaming system. I put on special contact lenses that give me full 360-degree views and impeccable fictional apparitions. With the contacts on, it is very difficult to visually ascertain if what I see is fake—except that one part of my brain is aware that a seven-meter-tall Godzilla stalking the street is absolute fantasy. I wear a ring on one finger of each hand to track my gestures. Tiny lenses in my shirt and headband track my body orientation. And GPS in my pocket device tracks my location to within a few millimeters. I can thus wander through my hometown as if it were an alternative world or a game platform. When I rush through the real streets, ordinary objects and spaces are transformed into extraordinary objects and spaces. A real newspaper rack on the real sidewalk becomes an elaborate 22nd-century antigravity transponder in an AR game.

The most intense VR experience of all requires a full-body VR rig. It's a lot of trouble so I suit up only occasionally. I have an amateur rig at home that includes a standing harness to prevent me from falling while I flail about. It gives me a full cardio workout while chasing dragons. In fact, VR harnesses have replaced exercise equipment in most basements. But once or twice a month I join some friends at the local realie theater to get access to state-of-the-art VR technology. Wearing my own silk underwear suit for hygienic purposes, I slip into an inflatable exoskeleton that closes around my limbs. This generates amazing haptic feedback. When I grasp a virtual object with my virtual hand, I feel its weight—the pressure against my hand—because the inflatable is squeezing my hand just the right amount. If I bump my shin against a rock in the virtual world, the sheath on my leg will "bump" my shin just so, making a totally believable sensation. A reclining seat holds my torso, giving me the option of doing genuinely felt jumps, flips, and dashes. And the accuracy of the super-hi-res helmet, with binaural sound and even real-time smells, creates a totally convincing presence. Within two minutes of entering, I usually forget where my real body is; I am elsewhere. The best part of a realie theater is that with zero latency 250 other people are sharing my world with equal verisimilitude. With them I can do real things in a fantasy world.

———

VR technology offers one more benefit to users. The strong presence generated by VR amplifies two paradoxically opposing traits. It enhances realness, so we might regard a fake world as real—the goal of many games and movies. And it encourages unrealness, fakery to the nth degree. For instance, it is easy to tweak the physics in VR to, say, remove gravity or friction, or to model fictional environments simulating alien planets— say, an underwater civilization. We can also alter our avatars to become other genders, other colors, or other species. For 25 years Jaron Lanier has talked about his desire to use VR to turn himself into a walking lobster. The software would swap his arms for claws, his ears for antennae, and his feet for a tail, not just visually, but kinetically. Recently at the Stanford VR lab Lanier's dream came true. VR creation software is now agile and robust enough to quickly model such personal fantasies. Using the Stanford VR rig, I too got to modify my avatar. In the experiment, once I was in VR, my arms would become my feet, and my feet my arms. That is, to kick with my virtual foot I had to punch with my real arm. To test how well this inversion worked, I had to burst floating virtual balloons with my arms/feet and feet/arms. The first seconds were awkward and embarrassing. But amazingly, within a few minutes I could kick with my arms and punch with my feet. Jeremy Bailenson, the Stanford professor who devised this experiment and uses VR as the ultimate sociological lab, discovered that it usually took a person only four minutes to completely rewire the feet/arm circuits in their brain. Our identities are far more fluid than we think.

That's becoming a problem. It's very difficult to determine how real someone online is. Outward appearances are easily manipulated. Someone may present himself as a lobster, but in reality he is a dreadlocked computer engineer. Formerly you could check their friends to ascertain realness. If a person online did not have any friends on social networks, they probably weren't who they claimed to be. But now hackers/criminals/rebels can create puppet accounts, with imaginary friends and imaginary friends of friends, working for bogus companies with bogus Wikipedia entries. The most valuable asset that Facebook owns is not its software platform but the fact that it controls the "true name" identities of a billion people, which are verified from references of the true identities of friends and colleagues. That monopoly of a persistent identity is the real engine of Facebook's remarkable success. And it is fragile. The normal tests we used to prove who we are in digital worlds, such as passwords and captchas, no long work very well. A captcha is a visual puzzle that was easy for humans to solve, but hard for computers. Now humans have trouble solving them, while machines find it easier. Passwords are easily hacked or stolen. So what is the better solution than passwords? You, yourself.

Your body is your password. Your digital identity is you. All the tools that VR is exploiting, all the ways it needs to capture your movements, to follow your eyes, to decipher your emotions, to encapsulate you as much as possible so you can be transported into another realm and believe you were there—all these interactions will be unique to you, and therefore proof of you. One of the recurring surprises in the field of biometrics— the science behind the sensors that track your body—is that almost everything that we can measure has a personally unique fingerprint. Your heartbeat is unique. Your gait when you walk is unique. Your typing rhythm on a keyboard is distinctive. What words you use most frequently. How you sit. Your blinks. Of course, your voice. When these are combined, they fuse into a metapattern that almost can't be faked. Indeed, that's how we identify people in the real world. If I were to meet you and was asked if we had met before, my subconscious mind would churn through a spectrum of subtle attributes—voice, face, body, style, mannerisms, bearing—before aggregating them into a recognition or not. In the technological world, we'll come to inspect a person with nearly the same spectrum of metrics. The system will check out a candidate's attributes. Do the pulse, breathing, heart rate, voice, face, iris, expressions, and dozens of other imperceptible biological signatures match who (or what) they claim? Our interactions will become our password.

Degrees of interaction are rising, and will continue to increase. Yet simple noninteractive things, such as a wooden-handled hammer, will endure. Still, anything that can interact, including a smart hammer, will become more valuable in our interactive society. But high interactivity comes at a cost. Interacting demands skills, coordination, experience, and education. Embedded into our technology and cultivated in ourselves. All the more so because we have only begun to invent novel ways to interact. The future of technology resides, in large part, in the discovery of new interactions. In the coming 30 years, anything that is not intensely interactive will be considered broken.

10

TRACKING

We are opaque to ourselves and need all the help we can get to decipher who we are. One modern aid is self-measurement. But the noble pursuit of unmasking our hidden nature with self-measurement has a short history. Until recently it took an especially dedicated person to find a way to measure themselves without fooling themselves. Scientific self-tracking was expensive, troublesome, and limited. But in the last few years extremely tiny digital sensors that cost a few pennies have made recording parameters so easy (just click a button), and the varieties of parameters so vast, that almost anyone can now measure a thousand different aspects of themselves. Already these self-experiments have started to change our ideas of medicine, health, and human behavior.

Digital magic has shrunk devices such as thermometers, heart rate monitors, motion trackers, brain wave detectors, and hundreds of other complex medical appliances to the size of words on this page. A few are shrinking to the size of the period following this sentence. These macroscopic measurers can be inserted into watches, clothes, spectacles, or phones, or inexpensively dispersed in our rooms, cars, offices, and public spaces.

In the spring of 2007 I was hiking with Alan Greene, a doctor friend of mine, in the overgrown hills behind my house in northern California. As we slowly climbed up the dirt path to the summit, we discussed a recent innovation: a tiny electronic pedometer that slipped into the laces of a shoe to record each step, then saved the data to an iPod for later analysis. We could use this tiny device to count the calories as we climbed and to track our exercise patterns over time. We began to catalog other available ways to measure our activities. A week later, I took the same hike with Gary Wolf, a writer for *Wired* magazine, who was curious about the social implications of these emerging self-tracking devices. There were only a dozen existing ones, but we both could see clearly that tracking technology would explode as sensors steadily got smaller. What to call this cultural drift? Gary pointed out that by relying on numbers instead of words we were constructing a "quantified self." So in June 2007 Gary and I announced on the internets that we would host a "Quantified Self" Meetup, open to absolutely anyone who thought they were quantifying themselves. We left the definition wide open to see who would show up. More than 20 people arrived at my studio in Pacifica, California, for this first event.

The diversity of what they were tracking astounded us: They measured their diet, fitness, sleep patterns, moods, blood factors, genes, location, and so on in quantifiable units. Some were making their own devices. One guy had been self-tracking for five years in order to maximize his strength, stamina, concentration, and productivity. He was using self-tracking in ways we had not imagined. Today there are 200 Quantified Self Meetup groups around the world, with 50,000 members. And every month, without fail, for eight years, someone at a Quantified Self meeting has demo'd an ingenious new way to track an aspect of their life that seemed unlikely or impossible a moment before. A few individuals stand out for their extreme habits. But what seems extreme today will soon become the new normal.

Computer scientist Larry Smarr tracks about a hundred health parameters on a daily basis, including his skin temperature and galvanic skin response. Every month he sequences the microbial makeup of his excrement, which mirrors the makeup of his gut microfauna, which is fast becoming one of the most promising frontiers in medicine. Equipped with this flow of data, and with a massive amount of amateur medical sleuthing, Smarr self-diagnosed the onset of Crohn's disease, or ulcerative colitis, in his own body, before he or his doctors noticed any symptoms. Surgery later confirmed his self-tracking.

Stephen Wolfram is the genius behind Mathematica, a clever software app that is a math processor (instead of a word processor). Being a numbers guy, Wolfram applied his numeracy to the 1.7 million files he archived about his life. He processed all his outgoing and incoming email for 25 years. He captured every keystroke for 13 years, logged all his phone calls, his steps, his room-to-room motion in his home/office, and his GPS location outside his house. He tracked how many edits he made while writing his books and papers. Using his own Mathematica program, he turned his self-tracking into a "personal analytics" engine, which illuminated patterns in his routines over several decades. Some patterns were subtle enough, such as the hours when he is most productive, that he had not detected them until he analyzed his own data.

Nicholas Felton is a designer who has also tracked and analyzed all of his emails, messages, Facebook and Twitter postings, phone calls, and travel for the past five years. Every year he generates an annual report in which he visualizes the previous year's data findings. In 2013 he concluded that he was productive on average 49 percent of the time, but most productive on Wednesdays, when he was 57 percent productive. At any given moment there is a 43 percent chance he is alone. He spent a third of his life (32 percent) sleeping. He used this quantitative review to help him "do a better job," including remembering the names of people he met.

At Quantified Self meetings we've heard from people who track their habitual tardiness, or the amount of coffee they drink, their alertness, or the number of times they sneeze. I can honestly say that anything that can be tracked is being tracked by someone somewhere. At a recent international Quantified Self conference, I made this challenge: Let's think of the most unlikely metric we can come up with and see if someone is tracking it. So I asked a group of 500 self-trackers: Is anyone tracking their fingernail growth? That seemed pretty absurd. One person raised their hand.

Shrinking chips, stronger batteries, and cloud connectivity has encouraged some self-trackers to attempt very long-term tracking. Particularly of one's health. Most people are lucky to see a doctor once a year to get some aspect of their health measured. But instead of once a year, imagine that every day, all day, invisible sensors measured and recorded your heart rate, blood pressure, temperature, glucose, blood serum, sleep patterns, body fat, activity levels, mood, EKG brain functions, and so on. You would have hundreds of thousands of data points for each of these traits. You would have evidence while at both rest and at full stress, while sick and healthy, in all seasons, all conditions. Over the years you would gain a very accurate measurement of your normal—the narrow range your levels meander in. It turns out that, in medicine, normal is a fictional average. Your normal is not my normal and vice versa. The average normal is not very useful to you specifically. But with long-term self-tracking, you'd arrive at a very personal baseline—*your normal*—which becomes invaluable when you are not feeling well, or when you want to experiment.

The achievable dream in the near future is to use this very personal database of your body's record (including your full sequence of genes) to construct personal treatments and personalized medicines. Science would use your life's log to generate treatments specifically for you. For instance, a smart personalized pill-making machine in your home (described in Chapter 7) would compound medicines in the exact proportions for your current bodily need. If the treatment in the morning eased the symptoms, the dosage in the evening would be adjusted by the system.

The standard way of doing medical research today is to run experiments on as many subjects as one possibly can. The higher the number (N) of subjects, the better. An N of 100,000 random people would be the most accurate way to extrapolate results to the entire population of the country because the inevitable oddballs within the test population would average out and disappear from the results. In fact, the majority of medical trials are conducted with 500 or fewer participants for economic reasons. But a scientific study where N=500, if done with care, can be good enough for an FDA drug approval.

A quantified-self experiment, on the other hand, is just N=1. The subject is yourself. At first it may seem that an N=1 experiment is not scientifically valid, but it turns out that it is extremely valid to *you*. In many ways it is the ideal experiment because you are testing the variable X against the very particular subject that is your body and mind at one point in time. Who cares whether the treatment works on anyone else? What you want to know is, How does it affect *me*? An N=1 provides that laser-focused result.

The problem with an N=1 experiment (which was once standard procedure for all medicine before the age of science) is not that the results aren't useful (they are), but that it is very easy to fool yourself. We all have hunches and expectations about our bodies, or about things we eat, or ideas of how the world works (such as the theory of vapors, or vibrations, or germs), that can seriously blind us to what is really happening. We suspect malaria is due to bad air, so we move to higher ground, and that helps, a little. We suspect gluten is giving us bloat, and so we tend to find evidence in our lives that it is the culprit and then we ignore contrary evidence that it doesn't matter. We are particularly susceptible to bias when we are hurting or desperate. An N=1 experiment can work only if we can separate the ordinary expectations of the experimenter from those of the subject, but since one person plays both roles, this is extremely hard. This kind of inbred prejudice is exactly what large randomized double-blind trials were invented to overcome. The subject is unaware of the parameters of the test and therefore cannot be biased. What helps overcome some of our self-fooling in an N=1 experiment in the new era of self-tracking is automatic instrumentation (having a sensor make the measurement many times for long periods so it is "forgotten" by the subject) and being able to track many variables at once to distract the subject, and then using statistical means later to try to unravel any patterns.

We know from many classic large population studies that often the medicine we take works because we believe it will work. This is otherwise known as the placebo effect. These quantified-self tricks don't fully counter the placebo effect; rather they work with it. If the intervention is producing a measurable improvement in *you*, then it works. Whether this measurable improvement is caused by the placebo effect doesn't matter since we only care what effect it has on this N=1 subject. Thus a placebo effect can be positive.

In formal studies, you need a control group to offset your bias toward positive results. So in lieu of a control group in an N=1 study, a quantified-self experimenter uses his or her own baseline. If you track yourself long enough, with a wide variety of metrics, then you can establish your behavior outside (or before) the experiment, which effectively functions as the control for comparison.

———

All this talk about numbers hides an important fact about humans: We have lousy mathematical intuitions. Our brains don't do statistics well. Math is not our natural language. Even extremely visual plots and numerical graphs demand superconcentration. In the long term, the quantification in the quantified self will become invisible. Self-tracking will go far beyond numbers.

Let me give you an example. In 2004, Udo Wachter, an IT manager in Germany, took the guts of a small digital compass and soldered it into a leather belt. He added 13 miniature piezoelectric vibrators, like the ones that vibrate your smartphone, and buried them along the length of the belt. Finally he hacked the electronic compass so that instead of displaying north on a circular screen, it vibrated different parts of the belt when it was clasped into a circle. The section of the circle "facing" north would always vibrate. When Udo put the belt on, he could feel northness on his waist. Within a week of always wearing the north belt, Udo had an unerring sensation of "north." It was unconscious. He could point in the direction without thinking. He just knew. After several weeks he acquired an additional heightened sense of location, of where he was in a city, as if he could feel a map. Here the quantification from digital tracking was subsumed into a wholly new bodily sensation. In the long term this is the destiny of many of the constant streams of data flowing from our bodily sensors. They won't be numbers; they will be new senses.

These new synthetic senses are more than entertaining. Our natural senses evolved over millions of years to ensure that we survived in a world of scarcity. The threat of not having enough calories, salt, or fat was relentless. As Malthus and Darwin showed, every biological population expands right to the limit of its starvation. Today, in a world made abundant by technology, the threat to survival is due to an excess of good stuff. Too much goodness throws our metabolism and psychology out of kilter. But our bodies can't register these new imbalances very well. We didn't evolve to sense our blood pressure or glucose levels. But our technology can. For instance, a new self-tracking device, the Scout from Scanadu, is the size of an old-timey stopwatch. By touching it to your forehead, it will measure your blood pressure, variable heart rate, heart performance (ECG), oxygen level, temperature, and skin conductance all in a single instant. Someday it will also measure your glucose levels. More than one startup in Silicon Valley is developing a noninvasive, prickless blood monitor to analyze your blood factors daily. You'll eventually wear these. By taking this information and feeding it back not in numbers but in a form we can feel, such as a vibration on our wrist or a squeeze on our hip, the device will equip us with a new sense about our bodies that we didn't evolve but desperately need.

———

Self-tracking is much broader than health. It is as big as our life itself. Tiny wearable digital eyes and ears can record every second of our entire day—who we saw and what we said—to aid our memories. Our stream of email and text, when saved, forms an ongoing diary of our mind. We can add the record of the music we listened to, the books and articles we read, the places we visited. The significant particulars of our routine movements and meetings, as well as nonroutine events and experiences, can also be funneled into bits and merged into a chronological flow.

This flow is called a lifestream. First described by the computer scientist David Gelernter in 1999, a lifestream is more than just a data archive. Gelernter conceived of lifestreams as a new organizing interface for computers. Instead of an old desktop, a new chronological stream. Instead of a web browser, a stream browser. Gelernter and his graduate student Eric Freeman define the lifestream architecture like this:

> A lifestream is a time-ordered stream of documents that functions as a diary of your electronic life; every document you create and every document other people send you is stored in your lifestream. The tail of your stream contains documents from the past (starting with your electronic birth certificate). Moving away from the tail and toward the present, your stream contains more recent documents—pictures, correspondence, bills, movies, voice mail, software. Moving beyond the present and into the future,

the stream contains documents you will need: reminders, calendar items, to-do lists.

You can sit back and watch new documents arrive: they're plunked down at the head of the stream. You browse the stream by running your cursor down it—touch a document in the display and a page pops out far enough for you to glance at its contents. You can go back in time or go to the future and see what you're supposed to be doing next week or next decade. Your entire cyberlife is right there in front of you.

Every person generates their own lifestream. When I meet with you, your lifestream and mine intersect in time. If we are going to meet next week, they intersect in the future; if we met, or even shared a photo last year, then our lifestreams intersected in the past. Our streams become richly braided with incredible complexity, but the strict chronological nature of each one means that they are easy to navigate. We naturally slide along a timeline to home in on an event. "It happened after the Christmas trip but before my birthday."

The advantage of a lifestream as an organizational metaphor, Gelernter says, is that "the question 'Where did I put that piece of information?' always has exactly one answer: It's in my stream. The idea of a timeline, a chronology, a diary, a daily journal, or a scrapbook is so much older and so much more organic and ingrained in human culture and history than the idea of a file hierarchy." As Gelernter told a Sun computer representative, "When I acquire a new memory of (let's say) talking to Melissa on a sunny afternoon outside the Red Parrot—I don't have to give this memory a name, or stuff it in a directory. I can use anything in the memory as a retrieval key. I shouldn't have to name electronic documents either, or put them in directories. I can shuffle other streams into mine—to the extent I have permission to use other people's streams. My own personal stream, my electronic life story, can have other streams shuffled into it—streams belonging to groups or organizations I'm part of. And eventually I'll have, for example, newspaper and magazine streams shuffled into my stream also."

Gelernter tried many times since 1999 to produce a commercial version of his software, but it never took off. A company that bought his patents sued Apple for stealing his Lifestream idea and using it in its Time Machine backup system. (To restore a file in Apple's Time Machine, you slide along a timeline to the date you want and there is "snapshot" of your computer's content on that date.)

But in social media today we have several working examples of lifestreams: Facebook (and in China, WeChat). Your Facebook stream is an ongoing flow of pictures, updates, links, pointers, and other documentation from your life. New pieces are continually added to the front of the stream. If you care to, you can add widgets to Facebook that capture the music you are listening to or the movies you are streaming. Facebook even provides a timeline interface to review the past. Over a billion other people's streams can intersect with yours. When a friend (or stranger) likes a post or tags a person in a picture, those two streams mingle. And each day Facebook is adding more current events and news streams and company updates into the worldstream.

But even all this is still only part of the picture. Lifestreaming can be thought of as an active, conscious tracking. People actively curate their stream when they snap a photo on their phones, or tag friends, or deliberately check-in to a place with Foursquare. Even their exercise Fitbit data, counting steps, is active, in that it is meant to be paid attention to. You can't change your behavior unless you pay attention in some capacity.

There is an equally important domain of tracking that is not conscious or active. This passive type of tracking is sometimes called lifelogging. The idea is to simply, mechanically, automatically, mindlessly, completely track everything all the time. Record everything that is recordable without prejudice, and for all your life. You only pay attention to it in the future when you may need it. Lifelogging is a hugely wasteful and inefficient process since most of what you lifelog is never used. But like many inefficient processes (such as evolution), it also contains genius. Lifelogging is possible now only because computation and storage and sensors have become so cheap that we can waste them with little cost. But creative "wasting" of computation has been the recipe for many of the most successful digital products and companies, and the benefits of lifelogging also lie in its extravagant use of computation.

Among the very first to lifelog was Ted Nelson in the mid-1980s (although he didn't call it that). Nelson, who invented hypertext, recorded every conversation he had with anyone on audio or videotape, no matter where or of what importance. He met and spoke to thousands of people, so he had a large rental storage container full of tapes. The second person was Steve Mann in the 1990s. Mann, then at MIT (now at the University of Toronto), outfitted himself with a head-mounted camera and recorded his daily life on videotape. Everything, all day, all year. For 25 years, if he was awake, he kept the camera on. His gear had a tiny screen over one eye and the camera recorded his first-person viewpoint, foreshadowing Google Glass by two decades. When we first met in July 1996, Mann sometimes called what he did "Quantimetric Self Sensing." Because there was a camera half obscuring his face, I found it was hard to be natural around Mann, but he is still routinely recording his whole life all the time.

But Gordon Bell at Microsoft Research may be the paragon of lifeloggers. For six years beginning in 2000, Bell documented every aspect of his work life in a grand experiment he called MyLifeBits. Bell wore a special custom-made camera around his neck that noticed a person's body heat if they were near and photographed them every 60 seconds. Bell's bodycam also snapped a picture if it detected a change in light of a new place. Bell recorded and archived every keystroke on his computer, every email, every website he visited, every search he made, every window on his computer and how long it remained opened. He also recorded many of his conversations, which enabled him to "scroll back" whenever there was disagreement on what had been said. He also scanned all his incoming pieces of paper into digital files and transcribed every phone conversation (with permission). Part of the intent of this experiment was to find out what kind of lifelogging tools Microsoft might want to invent to help workers manage the ocean of data this lifelogging generates—because making sense of all this data is a far bigger challenge than merely recording it.

The point of lifelogging is to create total recall. If a lifelog records everything in your life, then it could recover anything you experienced even if your meaty mind may have forgotten it. It would be like being able to google your life, if in fact your life were being indexed and fully saved. Our biological memories are so spotty that any compensation would be a huge win. Bell's experimental version of total recall helped increase his productivity. He could verify facts from previous conversations or recover insights he had forgotten. His system had little problem recording his life into bits, but he learned retrieving the meaningful bits needed better tools.

I've been wearing a tiny camera that I clip to my shirt, inspired by the one Gordon Bell wore. The Narrative is about an inch square. It takes a still photo every minute all day long, or whenever I wear it. I can also force a shot by tapping on the square twice. The photos go to the cloud, where they are processed and then sent back to my phone or the web. Narrative's software smartly groups the images into scenes during my day and then selects the most representative three images for each scene. This reduces the flood of images. Using this visual summary, I can flick through the 2,000 images per day very quickly, and then expand the stream of a particular scene for more images to find the exact moment I want to recall. I can easily browse the lifestream of an entire day in less than a minute. I find it mildly useful as a very detailed visual diary, a lifelogging asset that needs to be invaluable only a couple of times a month to make it worthwhile.

Typical users, Narrative has found, employ this photo diary while they attend conferences, or go on vacation, or want to record an experience. Recalling a conference is ideal. The continuous camera captures the many new people you meet. Better than a business card, you can much more easily recall them years later, and what they talked about, by browsing your lifestream. The photo lifestream is a strong prompt for vacations and family events. For instance, I recently used the Narrative during my nephew's wedding. It includes not only the iconic moments shared by everyone, but captured the conversations I had with people I had not talked to before. This version of Narrative does not record audio, but the next version will. In his research Bell discovered that the most informative media to capture is audio, prompted and indexed by photos. Bell told me that if he could have only one, he'd rather have an audio log of his day than a visual log.

An embrace of an expanded version of lifelogging would offer these four categories of benefits:

- **A constant 24/7/365 monitoring of vital body measurements.** Imagine how public health would change if we continuously monitored blood glucose in real time. Imagine how your behavior would change if you could, in near real time, detect the presence or absence of biochemicals or toxins in your blood picked up from your environment. (You might conclude: "I'm not going back there!") This data could serve both as a warning system and also as a personal base upon which to diagnose illness and prescribe medicines.
- **An interactive, extended memory of people you met, conversations you had, places you visited, and events you participated in.** This memory would be searchable, retrievable, and shareable.
- **A complete passive archive of everything that you have ever produced, wrote, or said.** Deep comparative analysis of your activities could assist your productivity and creativity.
- **A way of organizing, shaping, and "reading" your own life.**

To the degree this lifelog is shared, this archive of information could be leveraged to help others work and to amplify social interactions. In the health realm, shared medical logs could rapidly advance medical discoveries.

For many skeptics, there are two challenges that will doom lifelogging to a small minority. First, current social pressure casts self-tracking as the geekiest thing you could possibly do. Owners of Google Glass quickly put them away because they didn't like how they looked and they felt uncomfortable recording among their friends—or even uncomfortable explaining why they were not recording. As Gary Wolf said, "Recording in a diary is considered admirable. Recording in a spreadsheet is considered creepy." But I believe we'll quickly invent social norms and technological innovations to navigate the times when lifelogging is appropriate or not. When cell phones first appeared among the early adopters in the 1990s, there was a terrible cacophony of ringers. Cell phones rang at high decibels on trains, in bathrooms, in movie theaters. While talking on an early cell phone, people raised their voices as loud as the ringers. If you imagined back then what the world would sound like in the near future when everyone had a cell phone, you could only envision a nonstop racket. That didn't happen. Silent vibrators were invented, people learned to text, and social norms prevailed. I can go to a movie today in which every person in the theater has a cell phone, and not hear one ring or even see one lighted screen. It's considered not cool. We'll evolve the same kind of social conventions and technical fixes that will make lifelogging acceptable.

Second, how can lifelogging work when each person will generate petabytes, if not exabytes, of data each year? There is no way anyone can troll through that ocean of bits. You'll drown without a single insight. That is roughly true with today's software. Making sense of the data is an immense, time-consuming problem. You have to be highly numerate, technically agile, and supremely motivated to extract meaning from the river of data you generate. That is why self-tracking is still a minority sport. However, cheap artificial intelligence will overcome much of this. The AI in research labs is already powerful enough to sift through billions of records and surface important, meaningful patterns. As just one example, the same AI at Google that can already describe what is going on in a random photo could (when it is cheap enough) digest the images from my Narrative shirt cam so that I can simply ask Narrative in plain English to find me the guy who was wearing a pirate hat at a party I attended a couple of years ago. And there it is, and his stream would be linked to mine. Or I could ask it to determine the kind of rooms that tend to raise my heart rate. Was it the color, the temperature, the height of the ceilings? Although it seems like wizardry now, this will be considered a very mechanical request in a decade, not very different from asking Google to find something—which would have been magical 20 years ago.

Still, the picture is not big enough. We—the internet of people—will track ourselves, much of our lives. But the internet of *things* is much bigger, and billions of things will track themselves too. In the coming decades nearly every object that is manufactured will contain a small sliver of silicon that is connected to the internet. One consequence of this wide connection is that it will become feasible to track how each thing is used with great precision. For example, every car manufactured since 2006 contains a tiny OBD chip mounted under the dashboard. This chip records how your car is used. It tracks miles driven, at what speed, times of sudden braking, speed of turns, and gas mileage. This data was originally designed to help repair the car. Some insurance companies, such as Progressive, will lower your auto insurance rates if you give them access to your OBD driving log. Safer drivers pay less. The GPS location of cars can also be tracked very accurately, so it would be possible to tax drivers based on which roads they use and how often. These usage charges could be thought of as virtual tolls or automatic taxation.

———

The design of the internet of everything, and the nature of the cloud that it floats in, is to track data. The 34 billion internet-enabled devices we expect to add to the cloud in the next five years are built to stream data. And the cloud is built to keep the data. Anything touching this cloud that is able to be tracked will be tracked.

Recently, with the help of researcher Camille Hartsell, I rounded up all the devices and systems in the U.S. that routinely track us. The key word is "routinely." I am leaving off this list the nonroutine tracking performed illegally by hackers, criminals, and cyberarmies. I also skip over the capabilities of the governmental agencies to track specific targets when and how they want to. (Governments' ability to track is proportional to their budgets.) This list, instead, tallies the kind of tracking an average person might encounter on an ordinary day in the United States. Each example has been sourced officially or from a major publication.

Car movements—Every car since 2006 contains a chip that records your speed, braking, turns, mileage, accidents whenever you start your car.

Highway traffic—Cameras on poles and sensors buried in highways record the location of cars by license plates and fast-track badges. Seventy million plates are recorded each month.

Ride-share taxis—Uber, Lyft, and other decentralized rides record your trips.

Long-distance travel—Your travel itinerary for air flights and trains is recorded.

Drone surveillance—Along U.S. borders, Predator drones monitor and record outdoor activities.

Postal mail—The exterior of every piece of paper mail you send or receive is scanned and digitized.

Utilities—Your power and water usage patterns are kept by utilities. (Garbage is not cataloged, yet.)

Cell phone location and call logs—Where, when, and who you call (metadata) is stored for months. Some phone carriers routinely store the contents of calls and messages for days to years.

Civic cameras—Cameras record your activities 24/7 in most city downtowns in the U.S.

Commercial and private spaces—Today 68 percent of public employers, 59 percent of private employers, 98 percent of banks, 64 percent of public schools, and 16 percent of homeowners live or work under cameras.

Smart home—Smart thermostats (like Nest) detect your presence and behavior patterns and transmit these to the cloud. Smart electrical outlets (like Belkin) monitor power consumption and usage times shared to the cloud.

Home surveillance—Installed video cameras document your activity inside and outside the home, stored on cloud servers.

Interactive devices—Your voice commands and messages from phones (Siri, Now, Cortana), consoles (Kinect), smart TVs, and ambient microphones (Amazon Echo) are recorded and processed on the cloud.

Grocery loyalty cards—Supermarkets track which items you purchase and when.

E-retailers—Retailers like Amazon track not only what you purchase, but what you look at and even think about buying.

IRS—Tracks your financial situation all your life.

Credit cards—Of course, every purchase is tracked. Also mined deeply with sophisticated AI for patterns that reveal your personality, ethnicity, idiosyncrasies, politics, and preferences.

E-wallets and e-banks—Aggregators like Mint track your entire financial situation from loans, mortgages, and investments. Wallets like Square and PayPal track all purchases.

Photo face recognition—Facebook and Google can identify (tag) you in pictures taken by others posted on the web. The location of pictures can identify your location history.

Web activities—Web advertising cookies track your movements across the web. More than 80 percent of the top thousand sites employ web cookies that follow you wherever you go on the web. Through agreements with ad networks, even sites you did not visit can get information about your viewing history.

Social media—Can identify family members, friends, and friends of friends. Can identify and track your former employers and your current work mates. And how you spend your free time.

Search browsers—By default Google saves every question you've ever asked forever.

Streaming services—What movies (Netflix), music (Spotify), video (YouTube) you consume and when, and what you rate them. This includes cable companies; your watching history is recorded.

Book reading—Public libraries record your borrowings for about a month. Amazon records book purchases forever. Kindle monitors your reading patterns on ebooks—where you are in the book, how long you take to read each page, where you stop.

Fitness trackers—Your physical activity, time of day, sometimes location, often tracked all 24 hours, including when you sleep and when you are awake each day.

It is shockingly easy to imagine what power would accrue to any agency that could integrate all these streams. The fear of Big Brother stems directly from how technically easy it would be to stitch these together. At the moment, however, most of these streams are independent. Their bits are not integrated and correlated. A few strands may be coupled (credit cards and media usage, say), but by and large there is not a massive Big Brother–ish aggregate stream. Because they are slow, governments lag far behind what they could do technically. (Their own security is irresponsibly lax and decades behind the times.) Also, the U.S. government has not unified these streams because a thin wall of hard-won privacy laws holds them back. Few laws hold corporations back from integrating as much data as they can; therefore companies have become the proxy data gatherers for governments. Data about customers is the new gold in business, so one thing is certain: Companies (and indirectly governments) will collect more of it.

The movie *Minority Report*, based on a short story by Philip K. Dick, featured a not too distant future society that uses surveillance to arrest criminals before they commit a crime. Dick called that intervention "pre-crime" detection. I once thought Dick's idea of "pre-crime" to be utterly unrealistic. I don't anymore.

If you look at the above list of routine tracking today, it is not difficult to extrapolate another 50 years. All that was previously unmeasurable is becoming quantified, digitized, and trackable. We'll keep tracking ourselves, we'll keep tracking our friends, and our friends will track us. Companies and governments will track us more. Fifty years from now ubiquitous tracking will be the norm.

As I argue in Chapter 5 (Accessing), the internet is the world's largest, fastest copy machine, and anything that touches it will be copied. The internet wants to make copies. At first this fact is deeply troubling to creators, both individual and corporate, because their stuff will be copied indiscriminately, often for free, when it was once rare and precious. Some people fought, and still fight, very hard against the bias to copy (movie studios and music labels come to mind) and some people chose and choose to work with the bias. Those who embrace the internet's tendency to copy and seek value that can't be easily copied (through personalization, embodiment, authentication, etc.) tend to prosper, while those who deny, prohibit, and try to thwart the network's eagerness to copy are left behind to catch up later. Consumers, of course, love the promiscuous copies and feed the machine to claim their benefits.

This bias to copy is technological rather than merely social or cultural. It would be true in a different nation, even in a command economy, even with a different origin story, even on another planet. It is inevitable. But while we can't stop copying, it does matter greatly what legal and social regimes surround ubiquitous copying. How we handle rewards for innovation, intellectual property rights and responsibilities, ownership of and access to the copies makes a huge difference to society's prosperity and happiness. Ubiquitous copying is inevitable, but we have significant choices about its character.

Tracking follows a similar inevitable dynamic. Indeed, we can swap the term "tracking" in the preceding paragraphs for "copying" in the following paragraphs to get a sense of its parallels:

The internet is the world's largest, fastest tracking machine, and anything that touches it that can be tracked will be tracked. What the internet wants is to track everything. We will constantly self-track, track our friends, be tracked by friends, companies, and governments. This is deeply troubling to citizens, and to some extent to companies as well, because tracking was previously seen as rare and expensive. Some people fight hard against the bias to track and some will eventually work with the bias. Those who figure out how to domesticate tracking, to make it civil and productive, will prosper, while those who try only to prohibit and outlaw it will be left behind. Consumers say they don't want to be tracked, but in fact they keep feeding the machine with their data, because they want to claim their benefits.

This bias to track is technological rather than merely social or cultural. It would be true in a different nation, even in a command economy, even with a different origin story, even on another planet. But while we can't stop tracking, it does matter greatly what legal and social regimes surround it. Ubiquitous tracking is inevitable but we have significant choices about its character.

––––––

The fastest-increasing quantity on this planet is the amount of information we are generating. It is (and has been) expanding faster than anything else we can measure over the scale of decades. Information is accumulating faster than the rate we pour concrete (which is booming at a 7 percent increase annually), faster than the increases in the output of smartphones or microchips, faster than any by-product we generate, such as pollution or carbon dioxide.

Two economists at UC Berkeley tallied up the total global production information and calculated that new information is growing at 66 percent per year. This rate hardly seems astronomical compared with the 600 percent increase in iPods shipped in 2005. But that kind of burst is short-lived and not sustainable over decades (iPod production tanked in 2009). The growth of information has been steadily increasing at an insane rate for at least a century. It is no coincidence that 66 percent per year is the same as doubling every 18 months, which is the rate of Moore's Law. Five years ago humanity stored several hundred exabytes of information. That is the equivalent of each person on the planet having 80 Library of Alexandrias. Today we average 320 libraries each.

There's another way to visualize this growth: as an information explosion. Every second of every day we globally manufacture 6,000 square meters of information storage material—disks, chips, DVDs, paper, film—which we promptly fill up with data. That rate—6,000 square meters per second—is the approximate velocity of the shock wave radiating from an atomic explosion. Information is expanding at the rate of a nuclear explosion, but unlike a real atomic explosion, which lasts only seconds, this information explosion is perpetual, a nuclear blast lasting many decades.

In our everyday lives we generate far more information that we don't yet capture and record. Despite the explosion in tracking and storage, most of our day-to-day life is not digitized. This unaccounted-for information is "wild" or "dark" information. Taming this wild information will ensure that the total amount of information we collect will keep doubling for many decades ahead.

An increasing percentage of the information gathered each year is due to the information that we generate about that information. This is called meta-information. Every digital bit we capture encourages us to generate another bit concerning it. When the activity bracelet on my arm captures one step, it immediately adds time stamp data to it; it then creates yet more new data linking it to other step bits, and then generates tons of new data when it is plotted on a graph. Likewise, the musical data captured when a young girl plays her electric guitar on her live video stream becomes a foundation for generating indexing data about that clip, creating bits of data for "likes" or the many complex data packets needed to share that among her friends. The more data we capture, the more data we generate upon it. This metadata is growing even faster than the underlying information and is almost unlimited in its scale.

Metadata is the new wealth because the value of bits increases when they are linked to other bits. The least productive life for a bit is to remain naked and alone. A bit uncopied, unshared, unlinked with other bits will be a short-lived bit. The worst future for a bit is to be parked in some dark isolated data vault. What bits really want is to hang out with other related bits, be replicated widely, and maybe become a metabit, or an action bit in a piece of durable code. If we could personify bits, we'd say:

Bits want to move.

Bits want to be linked to other bits.

Bits want to be reckoned in real time.

Bits want to be duplicated, replicated, copied.

Bits want to be meta.

Of course, this is pure anthropomorphization. Bits don't have wills. But they do have tendencies. Bits that are related to other bits will tend to be copied more often. Just as selfish genes tend to replicate, bits do too. And just as genes "want" to code for bodies that help them replicate, selfish bits also "want" systems that help them replicate and spread. Bits behave as if they want to reproduce, move, and be shared. If you rely on bits for anything, this is good to know.

Since bits want to duplicate, replicate, and be linked, there's no stopping the explosion of information and the science fiction levels of tracking. Too many of the benefits we humans covet derive from streams of data. Our central choice now is: What kind of total tracking do we want? Do we want a one-way panopticon, where "they" know about us but we know nothing about them? Or could we construct a mutual, transparent kind of "coveillance" that involves watching the watchers? The first option is hell, the second tractable.

Not too long ago, small towns were the norm. The lady across the street from you tracked your every coming and going. She peeked out through her window and watched when you went to the doctor, and saw that you brought home a new TV, and knew who stayed with you over the weekend. But you also watched her through your window. You knew what she did on Thursday nights, and down at the corner drugstore you saw what she put in her basket. And there were mutual benefits from this mutual surveillance. If someone she did not recognize walked into your house when you were gone, she called the cops. And when she was gone, you picked up her mail from her mailbox. This small-town coveillance worked because it was symmetrical. You knew who was watching you. You knew what they did with the information. You could hold them accountable for its accuracy and use. And you got benefits for being watched. Finally, you watched your watchers under the same circumstances.

We tend to be uncomfortable being tracked today because we don't know much about who is watching us. We don't know what they know. We have no say in how the information is used. They are not accountable to correct it. They are filming us but we can't film them. And the benefits for being watched are murky and concealed. The relationship is unbalanced and asymmetrical.

Ubiquitous surveillance is inevitable. Since we cannot stop the system from tracking, we can only make the relationships more symmetrical. It's a way of civilizing coveillance. This will take both technological fixes and new social norms. Science fiction author David Brin calls this the "Transparent Society," which is also the name of his 1999 book summing up the idea. For a hint of how this scenario may be possible, consider Bitcoin, the decentralized open source currency described in Chapter 6 (Sharing). Bitcoin transparently logs every transaction in its economy in a public ledger, thereby making all financial transactions public. The validity of a transaction is verified by a coveillance of other users rather than the surveillance of central bank. For another example, traditional encryption used secret proprietary codes guarded closely. But a clever improvement called public key encryption (such as PGP) relies on code that anyone can inspect, including a public key, and therefore anyone can trust and verify. Neither of these innovations remedy existing asymmetries of knowledge; rather they demonstrate how it is possible to engineer systems that are powered by mutual vigilance.

In a coveillant society a sense of entitlement can emerge: Every person has a human right to access, and a right to benefit from, the data about themselves. But every right requires a duty, so every person has a human duty to respect the integrity of information, to share it responsibly, and to be watched by the watched.

The alternatives to coveillance are not promising. Outlawing the expansion of easy tracking will probably be as ineffectual as outlawing easy copying. I am a supporter of the whistle-blower Edward Snowden, who leaked tens of thousands of classified NSA files, revealing their role in secretly tracking citizens, primarily because I think the big sin of many governments, including the U.S., is lying about their tracking. Big governments are tracking us, but with no chance for symmetry. I applaud Snowden's whistle-blowing not because I believe it will reduce tracking, but because it can increase transparency. If symmetry can be restored so we can track who is tracking, if we can hold the trackers accountable by law (there should be regulation) and responsible for accuracy, and if we can make the benefits obvious and relevant, then I suspect the expansion of tracking will be accepted.

I want my friends to treat me as an individual. To enable that kind of relationship I have to be open and transparent and share my life with my friends so they know enough about me to treat me personally. I want companies to treat me as an individual too, so I have be open, transparent, and sharing with them as well to enable them to be personal. I want my government to treat me as an individual, so I have to reveal personal information to it to be treated personally. There is a one-to-one correspondence between personalization and transparency. Greater personalization requires greater transparency. Absolute personalization (vanity) requires absolute transparency (no privacy). If I prefer to remain private and opaque to potential friends and institutions, then I must accept I will be treated generically, without regard to my specific particulars. I'll be an average number.

Now imagine these choices pinned on a slider bar. On the left side of the slot is the pair *personal/transparent*. On the right side is the pair *private/generic*. The slider can slide to either side or anywhere in between. The slider is an important choice we have. Much to everyone's surprise, though, when technology gives us a choice (and it is vital that it remain a choice), people tend to push the slider all the way over to the *personal/transparent* side. They'll take transparent personalized sharing. No psychologist would have predicted that 20 years ago. If today's social media has taught us anything about ourselves as a species, it is that the human impulse to share overwhelms the human impulse for privacy. This has surprised the experts. So far, at every juncture that offers a choice, we've tilted, on average, toward more sharing, more disclosure, more transparency. I would sum it up like this: Vanity trumps privacy.

For eons and eons humans have lived in tribes and clans where every act was open and visible and there were no secrets. Our minds evolved with constant co-monitoring. Evolutionarily speaking, coveillance is our natural state. I believe that, contrary to our modern suspicions, there won't be a backlash against a circular world in which we constantly track each other because humans have lived like this for a million years, and—if truly equitable and symmetrical—it can feel comfortable.

That's a big if. Obviously, the relation between me and Google, or between me and the government, is inherently not equitable or symmetrical. The very fact they have access to everyone's lifestream, while I have access only to mine, means they have access to a qualitatively greater thing. But if some symmetry can be restored so that I can be part of holding their greater status to a greater accountability, and I benefit from their greater view, it might work. Put it this way: For sure cops will videotape citizens. That's okay as long as citizens can videotape cops, and can get access to the cops' videos, and share them to keep the more powerful accountable. That's not the end of the story, but it's how a transparent society has to start.

What about that state we used to call privacy? In a mutually transparent society, is there room for anonymity?

The internet makes true anonymity more possible today than ever before. At the same time the internet makes true anonymity in physical life much harder. For every step that masks us, we move two steps toward totally transparent unmasking. We have caller ID, but also caller ID block, and then caller ID–only filters. Coming up: biometric monitoring (iris + fingerprint + voice + face + heat rhythm) and little place to hide. A world where everything about a person can be found and archived is a world with no privacy. That's why many smart people are eager to maintain the option of easy anonymity—as a refuge for the private.

However, in every system that I have experienced where anonymity becomes common, the system fails. Communities saturated with anonymity will either self-destruct or shift from the purely anonymous to the pseudo-anonymous, as in eBay, where you have a traceable identity behind a persistent invented nickname. There is the famous outlaw gang Anonymous, an ad hoc rotating band of totally anonymous volunteers. They are online vigilantes with fickle targets. They will take down ISIS militant Twitter accounts, or a credit card company that gets in their way. But while they continue to persist and make trouble, it is not clear whether their net contribution to society is positive or negative.

For the civilized world, anonymity is like a rare earth metal. In larger doses these heavy metals are some of the most toxic substances known to a life. They kill. Yet these elements are also a necessary ingredient in keeping a cell alive. But the amount needed for health is a mere hard-to-measure trace. Anonymity is the same. As a trace element in vanishingly small doses, it's good, even essential for the system. Anonymity enables the occasional whistle-blower and can protect the persecuted fringe and political outcasts. But if anonymity is present in any significant quantity, it will poison the system. While anonymity can be used to protect heroes, it is far more commonly used as a way to escape responsibility. That's why most of the brutal harassment on Twitter, Yik Yak, Reddit, and other sites is delivered anonymously. A lack of responsibility unleashes the worst in us.

There's a dangerous idea that massive use of anonymity is a noble antidote to the prying state. This is like pumping up the level of heavy metals in your body to make it stronger. Rather, privacy can be gained only by trust, and trust requires persistent identity. In the end, the more trust the better, and the more responsibility the better. Like all trace elements, anonymity should never be eliminated completely, but it should be kept as close to zero as possible.

———

Everything else in the realm of data is headed to infinity. Or at least astronomical quantities. The average bit effectively becomes anonymous, almost undetectable, when measured against the scale of planetary data. In fact, we are running out of prefixes to indicate how big this new realm is. Gigabytes are on your phone. Terabytes were once unimaginably enormous, yet today I have three terabytes sitting on my desk. The next level up is peta. Petabytes are the new normal for companies. Exabytes are the current planetary scale. We'll probably reach zetta in a few years. Yotta is the last scientific term for which we have an official measure of magnitude. Bigger than yotta is blank. Until now, any more than a yotta was a fantasy not deserving an official name. But we'll be flinging around yottabytes in two decades or so. For anything beyond yotta, I propose we use the single term "zillion"—a flexible notation to cover any and all new magnitudes at this scale.

Large quantities of something can transform the nature of those somethings. More is different. Computer scientist J. Storrs Hall writes: "If there is enough of something, it is possible, indeed not unusual, for it to have properties not exhibited at all in small, isolated examples. There is no case in our experience where a difference of a factor of a trillion doesn't make a qualitative, as opposed to merely a quantitative, difference. A trillion is essentially the difference in weight between a dust mite, too small to see and too light to feel, and an elephant. It's the difference between $50 and a year's economic output for the entire human race. It's the difference between the thickness of a business card and the distance from here to the moon."

Call this difference zillionics.

A zillion neurons give you a smartness a million won't. A zillion data points will give you insight that a mere hundred thousand don't. A zillion chips connected to the internet create a pulsating, vibrating unity that 10 million chips can't. A zillion hyperlinks will give you information and behavior you could never expect from a hundred thousand links. The social web runs in the land of zillionics. Artificial intelligence, robotics, and virtual realities all require mastery of zillionics. But the skills needed to manage zillionics are daunting.

The usual tools for managing big data don't work very well in this territory. A statistical prediction technique such as a maximum likelihood estimation (MLE) breaks down because in the realm of zillionics the maximum likely estimate becomes improbable. Navigating zillions of bits, in real time, will require entire new fields of mathematics, completely new categories of software algorithms, and radically innovative hardware. What wide-open opportunities!

The coming new arrangement of data at the magnitude of zillionics promises a new machine at the scale of the planet. The atoms of this vast machine are bits. Bits can be arranged into complicated structures just as atoms are arranged into molecules. By raising the level of complexity, we elevate bits from data to information to knowledge. The full power of data lies in the many ways it can be reordered, restructured, reused, reimagined, remixed. Bits want to be linked; the more relationships a bit of data can join, the more powerful it gets.

The challenge is that the bulk of usable information today has been arranged in forms that only humans understand. Inside a snapshot taken on your phone is a long string of 50 million bits that are arranged in a way that makes sense to a human eye. This book you are reading is about 700,000 bits ordered into the structure of English grammar. But we are at our limits. Humans can no longer touch, let along process, zillions of bits. To exploit the full potential of the zillionbytes of data that we are harvesting and creating, we need to be able to arrange bits in ways that machines and artificial intelligences can understand. When self-tracking data can be cognified by machines, it will yield new, novel, and improved ways of seeing ourselves. In a few years, when AIs can understand movies, we'll be able to repurpose the zillionbytes of that visual information in entirely new ways. AI will parse images like we parse an article, and so it will be able to easily reorder image elements in the way we reorder words and phrases when we write.

Entirely new industries have sprung up in the last two decades based on the idea of unbundling. The music industry was overturned by technological startups that enabled melodies to be unbundled from songs and songs unbundled from albums. Revolutionary iTunes sold single songs, not albums. Once distilled and extracted from their former mixture, musical elements could be reordered into new compounds, such as shareable playlists. Big general-interest newspapers were unbundled into classifieds (Craigslist), stock quotes (Yahoo!), gossip (BuzzFeed), restaurant reviews (Yelp), and stories (the web) that stood and grew on their own. These new elements can be rearranged—remixed—into new text compounds, such as news updates tweeted by your friend. The next step is to unbundle classifieds, stories, and updates into even more elemental particles that can be rearranged in unexpected and unimaginable ways.

Sort of like smashing information into ever smaller subparticles that can be recombined into a new chemistry. Over the next 30 years, the great work will be parsing all the information we track and create—all the information of business, education, entertainment, science, sport, and social relations—into their most primeval elements. The scale of this undertaking requires massive cycles of cognition. Data scientists call this stage "machine readable" information, because it is AIs and not humans who will do this work in the zillions. When you hear a term like "big data," this is what it is about.

Out of this new chemistry of information will arise thousands of new compounds and informational building materials. Ceaseless tracking is inevitable, but it is only the start.

We are on our way to manufacturing 54 billion sensors every year by 2020. Spread around the globe, embedded in our cars, draped over our bodies, and watching us at home and on public streets, this web of sensors will generate another 300 zillionbytes of data in the next decade. Each of those bits will in turn generate twice as many metabits. Tracked, parsed, and cognified by utilitarian AIs, this vast ocean of informational atoms can be molded into hundreds of new forms, novel products, and innovative services. We will be astounded at what is possible by a new level of tracking ourselves.

11

QUESTIONING

Much of what I believed about human nature, and the nature of knowledge, was upended by Wikipedia. Wikipedia is now famous, but when it began I and many others considered it impossible. It's an online reference organized like an encyclopedia that unexpectedly allows anyone in the world to add to it, or change it, at any time, no permission needed. A 12-year-old in Jakarta could edit the entry for George Washington if she wanted to. I knew that the human propensity for mischief among the young and bored—many of whom lived online—would make an encyclopedia editable by anyone an impossibility. I also knew that even among the responsible contributors, the temptation to exaggerate and misremember was inescapable, adding to the impossibility of a reliable text. I knew from my own 20-year experience online that you could not rely on what you read by a random stranger, and I believed that an aggregation of random contributions would be a total mess. Even unedited web pages created by experts failed to impress me, so an entire encyclopedia written by unedited amateurs, not to mention ignoramuses, seemed destined to be junk.

Everything I knew about the structure of information convinced me that knowledge would not spontaneously emerge from data without a lot of energy and intelligence deliberately directed to transforming it. All the attempts at headless collective writing I had previously been involved with generated only forgettable trash. Why would anything online be any different?

So when the first incarnation of the online encyclopedia launched in 2000 (then called Nupedia), I gave it a look, and was not surprised that it never took off. While anyone could edit it, Nupedia required a laborious process of collaborative rewriting by other contributors that discouraged novice contributors. However, the founders of Nupedia created an easy-to-use wiki off to the side to facilitate working on the text, and much to everyone's surprise that wiki became the main event. Anyone could edit as well as post without waiting on others. I expected even less from that effort, now renamed Wikipedia.

How wrong I was. The success of Wikipedia keeps surpassing my expectations. At last count in 2015 it sported more than 35 million articles in 288 languages. It is quoted by the U.S. Supreme Court, relied on by schoolkids worldwide, and used by every journalist and lifelong learner for a quick education on something new. Despite the flaws of human nature, it keeps getting better. Both the weaknesses and virtues of individuals are transformed into common wealth, with a minimum of rules. Wikipedia works because it turns out that, with the right tools, it is easier to restore damaged text (the revert function on Wikipedia) than to create damaged text (vandalism), and so the good enough article prospers and continues to slowly improve. With the right tools, it turns out the collaborative community can outpace the same number of ambitious individuals competing.

It has always been clear that collectives amplify power—that is what cities and civilizations are—but what's been the big surprise for me is how minimal the tools and oversight that are needed. The bureaucracy of

Wikipedia is relatively so small as to be invisible, although it has grown over its first decade. Yet the greatest surprise brought by Wikipedia is that we still don't know how far this power can go. We haven't seen the limits of wiki-ized intelligence. Can it make textbooks, music, and movies? What about law and political governance?

Before we say, "Impossible!" I say: Let's see. I know all the reasons why law can never be written by know-nothing amateurs. But having already changed my mind once on this, I am slow to jump to conclusions again. A Wikipedia is impossible, but here it is. It is one of those things that is impossible in theory but possible in practice. Once you confront the fact that it works, you have to shift your expectation of what else there may be that is impossible in theory but might work in practice. To be honest, so far this open wiki model has been tried in a number of other publishing fields but has not been widely successful. Yet. Just as the first version of Wikipedia failed because the tools and processes were not right, collaborative textbooks, or law, or movies may take the invention of further new tools and methods.

I am not the only one who has had his mind changed about this. When you grow up having "always known" that such a thing as Wikipedia works, when it is obvious to you that open source software is better than polished proprietary goods, when you are certain that sharing your photos and other data yields more than safeguarding them—then these assumptions will become a platform for a yet more radical embrace of the common wealth. What once seemed impossible is now taken for granted.

Wikipedia has changed my mind in other ways. I was a fairly steady individualist, an American with libertarian leanings, and the success of Wikipedia led me toward a new appreciation of social power. I am now much more interested in both the power of the collective and the new obligations stemming from individuals toward the collective. In addition to expanding civil rights, I want to expand civil duties. I am convinced that the full impact of Wikipedia is still subterranean and that its mind-changing force is working subconsciously on the global millennial generation, providing them with an existent proof of a beneficial hive mind, and an appreciation for believing in the impossible.

More important, Wikipedia has taught me to believe in the impossible more often. In the past several decades I've had to accept other ideas that I formerly thought were impossibilities but that later turned out to be good practical ideas. For instance, I had my doubts about the online flea market called eBay when I first encountered it in 1997. You want me to transfer thousands of dollars to a distant stranger trying to sell me a used car I've never seen? Everything I had been taught about human nature suggested this could not work. Yet today, strangers selling automobiles is the major profit center for the very successful eBay corporation.

Twenty years ago I *might* have been able to believe that in 2016 we'd have maps for the entire world on our personal handheld devices. But I could not have been convinced we'd have them with street views of the buildings for many cities, or apps that showed the locations of public toilets, and that it would give us spoken directions for walking or public transit, and that we'd have all this mapping and more "for free." It seemed starkly impossible back then. And this free abundance still seems hard to believe in theory. Yet here it is on hundreds of millions of phones.

These supposed impossibilities keep happening with increased frequency. Everyone "knew" that people don't work for free, and if they did, they could not make something useful without a boss. But today entire sections of our economy run on software instruments created by volunteers working without pay or bosses. Everyone knew humans were innately private beings, yet the impossibility of total open round-the-clock sharing still occurred. Everyone knew that humans are basically lazy, and they would rather watch than create, and they would never get off their sofas to create their own TV. It would be impossible that millions of amateurs would produce billions of hours of video, or that anyone would watch any of it. Like Wikipedia, YouTube is theoretically impossible. But here again this impossibility is real in practice.

This list goes on, old impossibilities appearing as new possibilities daily. But why now? What is happening to disrupt the ancient impossible/possible boundary?

As far as I can tell, the impossible things happening now are in every case due to the emergence of a new level of organization that did not exist before. These incredible eruptions are the result of large-scale collaboration, and massive real-time social interacting, which in turn are enabled by omnipresent instant connection between billions of people at a planetary scale. Just as fleshy tissue yields a new, higher level of organization for a bunch of individual cells, these new social structures yield new tissue for individual humans. Tissue can do things that cells can't. The collectivist organizations of Wikipedia, Linux, Facebook, Uber, the web—even AI—can do things that industrialized humans could not. This is the first time on this planet that we've tied a billion people together in immediate syncopation, just as Facebook has done. From this new societal organization, new behaviors emerge that were impossible at the lower level.

Humans have long invented new social organizations, from law, courts, irrigation systems, schools, governments, libraries up to the largest scale, civilization itself. These social instruments are what makes us human—and what makes our behavior "impossible" from the vantage point of animals. For instance, when we invented written records and laws, these enabled a type of egalitarianism not possible in our cousins the primates, and not present in oral cultures. The cooperation and coordination bred by irrigation and agriculture produced yet more impossible behaviors of anticipation and preparation, and sensitivity to the future. Human society unleashed all kinds of previously impossible human behaviors into the biosphere.

The technium—the modern system of culture and technology—is

accelerating the creation of new impossibilities by continuing to invent new social organizations. The genius of eBay was its invention of cheap, easy, and quick reputation status. Strangers could sell to strangers at a great distance because we now had a technology to quickly assign persistent reputations to those beyond our circle. That lowly innovation opened up a new kind of higher-level coordination that permitted a new kind of exchange (remote purchasing among strangers) that was impossible before. The same kind of technologically enabled trust, plus real-time coordination, makes the decentralized taxi service Uber possible. The "revert log" button on Wikipedia, which made it easier to restore a vandalized passage than to vandalize it, unleashed a new higher organization of trust, emphasizing one facet of human behavior not enabled at a large scale before.

We have just begun to fiddle with social communications. Hyperlinks, wifi, and GPS location services are really types of relationships enabled by technology, and this class of innovations is just beginning. The majority of the most amazing communication inventions that are possible have not been invented yet. We are also just in the infancy of being able to invent institutions at a truly global scale. When we weave ourselves together into a global real-time society, former impossibilities will really start to erupt into reality. It is not necessary that we invent some kind of autonomous global consciousness. It is only necessary that we connect everyone to everyone else—and to everything else—all the time and create new things together. Hundreds of miracles that seem impossible today will be possible with this shared human connectivity.

I am looking forward to having my mind changed a lot in the coming years. I think we'll be surprised by how many of the things we assumed were "natural" for humans are not really natural at all. It might be fairer to say that what is natural for a tribe of mildly connected humans will not be natural for a planet of intensely connected humans. "Everyone knows" that humans are warlike, but I would guess organized war will become

less attractive, or useful, over time as new means of social conflict resolution arise at a global level. Of course, many of the impossible things we
can expect will be impossibly bad. The new technologies will unleash
whole new ways to lie, cheat, steal, spy, and terrorize. We have no consensual international rules for cyberconflict, which means we can expect
some very nasty unexpected "impossible" cyber events in the coming
decade. Because of our global connectivity, a relatively simple hack could
cause an emerging cascade of failure, which would reach impossible scale
very quickly. Worldwide disruptions of our social fabric are in fact inevitable. One day in the next three decades the entire internet/phone system
will blink off for 24 hours, and we'll be in shock for years afterward.

I don't focus on these expected downsides in this book for several
reasons. First, there is no invention that cannot be subverted in some way
to cause harm. Even the most angelic technology can be weaponized, and
will be. Criminals are some of the most creative innovators in the world.
And crap constitutes 80 percent of everything. But importantly, these
negative forms follow exactly the same general trends I've been outlining
for the positive. The negative, too, will become increasingly cognified,
remixed, and filtered. Crime, scams, warring, deceit, torture, corruption,
spam, pollution, greed, and other hurt will all become more decentralized and data centered. Both virtue and vice are subject to the same great
becoming and flowing forces. All the ways that startups and corporations
need to adjust to ubiquitous sharing and constant screening apply to
crime syndicates and hacker squads as well. Even the bad can't escape
these trends.

Additionally, it may seem counterintuitive, but every harmful invention also provides a niche to create a brand-new never-seen-before good.
Of course, that newly minted good can then be (and probably will be)
abused by a corresponding bad idea. It may seem that this circle of new
good provoking new bad which provokes new good which spawns new
bad is just spinning us in place, only faster and faster. That would be true

except for one vital difference: On each round we gain additional opportunities and choices that did not exist before. This expansion of choices (including the choice to do harm) is an increase in freedom—and this increase in freedoms and choices and opportunities is the foundation of our progress, of our humanity, and of our individual happiness.

Our technological spinning has thrown us up to a new level, opening up an entirely new continent of unknown opportunities and scary choices. The consequences of global-scale interactions are beyond us. The amount of data and power needed is inhuman; the vast realms of peta-, exa-, zetta-, zillion don't really mean anything to us today because this is the vocabulary of megamachines, and of planets. We will certainly behave differently collectively than as individuals, but we don't know how. Much more important, as individuals we behave differently in collectives.

This has been true for humans for a long while, ever since we moved to cities and began building civilizations. What's new now and in the coming decades is the velocity of this higher territory of connectivity (speed of light), and its immensely vaster scale (the entire planet). We are headed for a trillion times increase. As noted earlier, a shift by a trillion is not merely a change in quantity, but a change in essence. Most of what "everybody knows" about human beings has so far been based on the human individual. But there may be a million different ways to connect several billion people, and each way will reveal something new about us. Or each way may *create* in us something new. Either way, our humanity will shift.

Connected, in real time, in multiple ways, at an increasingly global scale, in matters large and small, with our permission, we will operate at a new level, and we won't cease surprising ourselves with impossible achievements. The impossibility of Wikipedia will quietly recede into outright obviousness.

In addition to hard-to-believe emergent phenomenon, we are headed to a world where the improbable is the new normal. Cops, emergency room doctors, and insurance agents see a bit of this already. They realize how many crazy impossible things actually happen all the time. For instance, a burglar gets stuck in a chimney; a truck driver in a head-on collision is thrown out his front window and lands on his feet, walking away; a wild antelope galloping across a bike trail knocks a man off his bicycle; a candle at a wedding ignites the bride's hair on fire; a girl casually fishing off a backyard dock catches a huge man-size shark. In former times these unlikely events would be private, known only as rumors, stories a friend of a friend told, easily doubted and not really believed.

But today they are on YouTube, and they fill our vision. You can see them yourself. Each of these weird freakish events has been seen by millions.

The improbable consists of more than just accidents. The internets are also brimming with improbable feats of performance—someone who can run up a side of a building, or slide down suburban rooftops on a snowboard, or stack up cups faster than you can blink. And not just humans—pets open doors, ride scooters, and paint pictures. The improbable also includes extraordinary levels of superhuman achievements: people doing astonishing memory tasks, or imitating all the accents of the world. In these extreme feats we see the super in humans.

Every minute a new impossible thing is uploaded to the internet and that improbable event becomes just one of hundreds of extraordinary events that we'll see or hear about today. The internet is like a lens that focuses the extraordinary into a beam, and that beam has become our illumination. It compresses the unlikely into a small viewable band of everydayness. As long as we are online—which is almost all day many days—we are illuminated by this compressed extraordinariness. It is the new normal.

That light of superness changes us. We no longer want mere presentations; we want the best, greatest, most extraordinary presenters alive, like in the TED videos. We don't want to watch people playing games; we want to watch the highlights of the highlights, the most amazing moves, catches, runs, shots, and kicks, each one more remarkable and improbable than the other.

We are also exposed to the greatest range of human experience: the heaviest person, shortest midgets, longest mustache—the entire universe of superlatives. Superlatives were once rare—by definition—but now we see multiple videos of superlatives all day long, and they seem normal. Humans have always treasured drawings and photos of the weird extremes of humanity (witness early issues of *National Geographic* and *Ripley's Believe It or Not*), but there is an intimacy about watching these extremities on our phones while we wait at the dentist. They are now much realer, and they fill our heads. I think there is already evidence that this ocean of extraordinariness is inspiring and daring ordinary folks to try something extraordinary.

At the same time, superlative epic failures are foremost as well. We are confronted by the stupidest people in the world doing the dumbest things imaginable. In some respects this may place us in a universe of nothing more than tiny, petty, obscure Guinness World Record holders. In every life there is probably at least one moment that is freakish, so everyone alive is a world record holder for 15 minutes. The good news may be that it cultivates in us an expanded sense of what is possible for humans, and for human life, and so extremism expands us. The bad news may be that this insatiable appetite for super-superlatives leads to dissatisfaction with anything ordinary.

There's no end to this dynamic. Cameras are ubiquitous, so as our collective tracked life expands, we'll accumulate thousands of videos showing people being struck by lightning—because improbable events are more normal than we think. When we all wear tiny cameras all the time, then the most improbable event, the most superlative achievement, the most extreme actions of anyone alive will be recorded and shared around the world in real time. Soon only the most extraordinary moments of 6 billion citizens will fill our streams. So henceforth rather than be surrounded by ordinariness we'll float in extraordinariness—as it becomes mundane. When the improbable dominates our field of vision to the point that it seems as if the world contains *only* the impossible, then these improbabilities don't feel as improbable. The impossible will feel inevitable.

There is a dreamlike quality to this state of improbability. Certainty itself is no longer as certain as it once was. When I am connected to the Screen of All Knowledge, to that billion-eyed hive of humanity woven together and mirrored on a billion pieces of glass, truth is harder to find. For every accepted piece of knowledge I come across, there is, within easy reach, a challenge to the fact. Every fact has its antifact. The internet's extreme hyperlinking will highlight those antifacts as brightly as the facts. Some antifacts are silly, some borderline, and some valid. This is the curse of the screen: You can't rely on experts to sort them out because for every expert there is an equal and opposite anti-expert. Thus anything I learn is subject to erosion by these ubiquitous antifactors.

Ironically, in an age of instant global connection, my certainty about anything has decreased. Rather than receiving truth from an authority, I am reduced to assembling my own certainty from the liquid stream of facts flowing through the web. Truth, with a capital T, becomes truths, plural. I have to sort the truths not just about things I care about, but about anything I touch, including areas about which I can't possibly have any direct knowledge. That means that in general I have to constantly question what I think I know. We might consider this state perfect for the advancement of science, but it also means that I am more likely to have my mind changed for incorrect reasons.

While hooked into the network of networks I feel like I am a network myself, trying to achieve reliability from unreliable parts. And in my quest to assemble truths from half-truths, nontruths, and some noble truths scattered in the flux, I find my mind attracted to fluid ways of thinking (scenarios, provisional belief, subjective hunches) and toward fluid media like mashups, twitterese, and search. But as I flow through this slippery web of ideas, it often feels like a waking dream.

We don't really know what dreams are for, only that they satisfy some fundamental need of consciousness. Someone watching me surf the web, as I jump from one suggested link to another, would see a daydream. On the web recently I found myself in a crowd of people watching a barefoot man eat dirt, then I saw a boy singing whose face began to melt, then Santa burned a Christmas tree, then I was floating inside a mud house on the very tippy top of the world, then Celtic knots untied themselves, then a guy told me the formula for making clear glass, then I was watching myself, back in high school, riding a bicycle. And that was just the first few minutes of my time surfing the web one morning. The trancelike state we fall into while following the undirected path of links could been seen as a terrible waste of time—or, like dreams, it might be a productive waste of time. Perhaps we are tapping into our collective unconscious as we roam the web. Maybe click-dreaming is a way for all of us to have the same dream, independent of what we click on.

This waking dream we call the internet also blurs the difference between my serious thoughts and my playful thoughts, or to put it more simply: I no longer can tell when I am working and when I am playing online. For some people the disintegration between these two realms marks all that is wrong with the internet: It is the high-priced waster of time. It breeds trifles and turns superficialities into careers. Jeff Hammerbacher, a former Facebook engineer, famously complained that the "best minds of my generation are thinking about how to make people click ads." This waking dream is viewed by some as an addictive squandering. On the contrary, I cherish a good wasting of time as a necessary precondition for creativity. More important, I believe the conflation of play and work, of thinking hard and thinking playfully, is one of the greatest things this new invention has done. Isn't the whole idea that in a highly evolved advanced society work is over?

I've noticed a different approach to my thinking now that the hive mind has spread it extremely wide and loose. My thinking is more active, less contemplative. Rather than begin a question or hunch by ruminating aimlessly in my mind, nourished only by my ignorance, I start doing things. I immediately *go*. I go looking, searching, asking, questioning, reacting, leaping in, constructing notes, bookmarks, a trail—I start off making something mine. I don't wait. Don't have to wait. I act on ideas first now instead of thinking on them. For some folks, this is the worst of the net—the loss of contemplation. Others feel that all this frothy activity is simply stupid busywork, or spinning of wheels, or illusionary action. But compared with what? Compared with the passive consumption of TV? Or time spent lounging at a bar chatting? Or the slow trudge to a library only to find no answers to the hundreds of questions I have? Picture the thousands of millions of people online at this very minute. To my eye they are not wasting time with silly associative links, but are engaged in a more productive way of thinking—getting instant answers, researching, responding, daydreaming, browsing, being confronted with something very different, writing down their own thoughts, posting their opinions, even if small. Compare that to the equivalent of hundreds of millions of people 50 years ago watching TV or reading a newspaper in a big chair.

This new mode of being—surfing the waves, diving down, rushing up, flitting from bit to bit, tweeting and twittering, ceaselessly dipping into newness with ease, daydreaming, questioning each and every fact—is not a bug. It is a feature. It is a proper response to the ocean of data, news, and facts flooding us. We need to be fluid and agile, flowing from idea to idea, because that fluidity reflects the turbulent informational environment surrounding us. This mode is neither a lazy failure nor an indulgent luxury. It is a necessity in order to thrive. To steer a kayak on white-water rapids you need to be paddling at least as fast as the water runs, and to hope to navigate the exabytes of information, change, disruption coming at us, you need to be flowing as fast as the frontier is flowing.

But don't confuse this flux for the shallows. Fluidity and interactivity also allow us to instantly divert more attention to works that are far more complex, bigger, and more complicated than ever before. Technologies that provided audiences with the ability to interact with stories and news— to time shift, play later, rewind, probe, link, save, clip, cut and paste— enabled long forms as well as short forms. Film directors started creating motion pictures that were not a series of sitcoms, but a massive sustained narrative that took years to tell. These vast epics, like *Lost, Battlestar Galactica, The Sopranos, Downton Abbey,* and *The Wire,* had multiple interweaving plotlines, multiple protagonists, and an incredible depth of characters, and these sophisticated works demanded sustained attention that was not only beyond previous TV and 90-minute movies, but would have shocked Dickens and other novelists of yore. Dickens would have marveled back then: "You mean the audience could follow all that, and then want more? Over how many years?" I would never have believed myself capable of enjoying such complicated stories, or caring about them enough to put in the time. My attention has grown. In a similar way the depth, complexity, and demands of video games can equal the demands of marathon movies or any great book. Just to become proficient in some games takes 50 hours.

But the most important way these new technologies are changing how we think is that they have become one thing. It may appear as if you are spending endless nanoseconds on a series of tweets, and infinite microseconds surfing between web pages, or hours wandering between YouTube channels, and then hovering only mere minutes on one book snippet after another, when you finally turn back to your spreadsheet at work or flick through the screen of your phone. But in reality you are spending 10 hours a day paying attention to one intangible thing. This one machine, this one huge platform, this gigantic masterpiece is disguised as a trillion loosely connected pieces. The unity is easy to miss. The well-paid directors of websites, the hordes of commenters online, and the movie moguls reluctantly letting us stream their movies—these folks don't believe they are mere data points in a big global show, but they are. When we enter any of the 4 billion screens lit today, we are participating in one open-ended question. We are all trying to answer: What is it?

The computer manufacturer Cisco estimates that there will be 50 billion devices on the internet by 2020, in addition to tens of billions of screens. The electronics industry expects a billion wearable devices in five years, tracking our activities, feeding data into the stream. We can expect another 13 billion appliances, like the Nest thermostat, animating our smarthomes. There will be 3 billion devices built into connected cars. And 100 billion dumb RFID chips embedded into goods on the shelves of Walmart. This is the internet of things, the emerging dreamland of everything we manufacture that is the new platform for the improbable. It is built with data.

Knowledge, which is related, but not identical, to information, is exploding at the same rate as information, doubling every two years. The number of scientific articles published each year has been accelerating even faster than this for decades. Over the last century the annual number of patent applications worldwide has risen in an exponential curve.

We know vastly more about the universe than we did a century ago. This new knowledge about the physical laws of the universe has been put to practical use in such consumer goods as GPS and iPods, with a steady increase in our own lifespans. Telescopes, microscopes, fluoroscopes, oscilloscopes allowed us to see in new ways, and when we looked with new tools, we suddenly gained many new answers.

Yet the paradox of science is that every answer breeds at least two new questions. More tools, more answers, ever more questions. Telescopes, radioscopes, cyclotrons, atom smashers expanded not only what we knew, but birthed new riddles and expanded what we didn't know. Previous discoveries helped us to recently realize that 96 percent of all matter and energy in our universe is outside of our vision. The universe is not made of the atoms and heat we discovered last century; instead it is primarily composed of two unknown entities we label "dark": dark energy and dark matter. "Dark" is a euphemism for ignorance. We really have no idea what the bulk of the universe is made of. We find a similar proportion of ignorance if we probe deeply into the cell, or the brain. We don't know nothin' relative to what could be known. Our inventions allow us to spy into our ignorance. If knowledge is growing exponentially because of scientific tools, then we should be quickly running out of puzzles. But instead we keep discovering greater unknowns.

Thus, even though our knowledge is expanding exponentially, our questions are expanding exponentially faster. And as mathematicians will tell you, the widening gap between two exponential curves is itself an exponential curve. That gap between questions and answers is our ignorance, and it is growing exponentially. In other words, science is a method that chiefly expands our ignorance rather than our knowledge.

We have no reason to expect this to reverse in the future. The more disruptive a technology or tool is, the more disruptive the questions it will breed. We can expect future technologies such as artificial intelligence, genetic manipulation, and quantum computing (to name a few on the near horizon) to unleash a barrage of new huge questions—questions we could have never thought to ask before. In fact, it's a safe bet that we have not asked our biggest questions yet.

––––––

Every year humans ask the internet 2 trillion questions, and every year the search engines give back 2 trillion answers. Most of those answers are pretty good. Many times the answers are amazing. And they are free! In the time before instant free internet search, the majority of the 2 trillion questions could not have been answered for any reasonable cost. Of course, while the answers may be free to users, they do cost the search companies like Google, Yahoo!, Bing, and Baidu something to create. In 2007, I calculated the cost to Google to answer one query to be approximately 0.3 cents, which has probably decreased a bit since then. By my calculations Google earns about 27 cents per search/answer from the ads placed around its answers, so it can easily afford to give its answers away for free.

We've always had questions. Thirty years ago the largest answering business was phone directory assistance. Before Google, there was 411. The universal "information" number 411 was dialed from phones about 6 billion times per year. The other search mechanism in the past was the yellow pages—the paper version. According to the Yellow Pages Association, 50 percent of American adults used the print yellow pages at least once a week, performing two lookups per week in the 1990s. Since the adult population in the 1990s was around 200 million, that's 200 million searches per week, or 104 billion questions asked per year. Nothing to sneeze at. The other classic answer strategy was the library. U.S. libraries in the 1990s counted about 1 billion library visits per year. Out of those 1 billion, about 300 million were "reference transactions," or questions.

Despite those 100 billion–plus searches for answers per year (in the U.S. alone), no one would have believed 30 years ago that there was an $82 billion business in answering people's questions for cheap or for free. There weren't many MBAs dreaming of schemes to fill this need. The demand for questions/answers was latent. People didn't know how valuable instant answers were until they had access to them. One study conducted in 2000 determined that the average American adult sought to answer four questions per day online. If my own life is any indication, I am asking more questions every day. Google told me that in 2007 I asked it 349 questions in one month, or 10 per day (and my peak hour of inquiry was 11 a.m. on Wednesdays). I asked Google how many seconds in a year and it instantly told me: 31.5 million. I asked it how many searches all search engines do per second? It said 600,000 searches per second, or 600 kilohertz. The internet is answering questions at the buzzing frequency of radio waves.

But while answers are provided for free, the value of those answers is huge. Three researchers at the University of Michigan performed a small experiment in 2010 to see if they could ascertain how much ordinary people might pay for search. Their method was to ask students inside a well-stocked university library to answer some questions that were asked on Google, but to find the answers only using the materials in the library. They measured how long it took the students to answer a question in the stacks. On average it took 22 minutes. That's 15 minutes longer than the 7 minutes it took to answer the same question, on average, using Google. Figuring a national average wage of $22 per hour, this works out to a savings of $1.37 per search.

In 2011, Hal Varian, the chief economist at Google, calculated the average value of answering a question in a different way. He revealed the surprising fact that the average user of Google (judged by returning cookies, etc.) makes only one search per day, on average. This is certainly not me. But my near constant googling is offset by, say, my mother, who may search only once every several weeks. Varian did some more math to compensate for the fact that because questions are now cheap we ask more of them. So when this effect is factored in, Varian calculated that search saves the average person 3.75 minutes per day. Using the same average hourly wage, people save 60 cents per day. We could even round that off to a dollar per day if your time is more valuable. Would most people pay a dollar per day, or $350 per year, for search if they had to? Maybe. (I absolutely would.) They might pay a dollar per search, which is another way of paying the same amount. Economist Michael Cox asked his students how much they would accept to give up the internet entirely and reported they would not give up the internet for a million dollars. And this was before smartphones became the norm.

We are just starting to get good at giving great answers. Siri, the audio phone assistant for the iPhone, delivers spoken answers when you ask her a question in natural English. I use Siri routinely. When I want to know the weather, I just ask, "Siri, what's the weather for tomorrow?" Android folks can audibly ask Google Now for information about their calendars. IBM's Watson proved that for most kinds of factual reference questions, an AI can find answers fast and accurately. Part of the increasing ease in providing answers lies in the fact that past questions answered correctly increase the likelihood of another question. At the same time, past correct answers increase the ease of creating the next answer, and increase the value of the corpus of answers as a whole. Each question we ask a search engine and each answer we accept as correct refines the intelligence of the process, increasing the engine's value for future questions. As we cognify more books and movies and the internet of things, answers become ubiquitous. We are headed to a future where we will ask several hundred questions per day. Most of these questions will concern us and our friends. "Where is Jenny? What time is the next bus? Is this kind of snack good?" The "manufacturing costs" of each answer will be nanocents. Search, as in "give me an answer," will no longer be considered a first-world luxury. It will become an essential universal commodity.

Very soon now we'll live in a world where we can ask the cloud, in conversational tones, any question at all. And if that question has a known answer, the machine will explain it to us. Who won the Rookie of the Year Award in 1974? Why is the sky blue? Will the universe keep expanding forever? Over time the cloud, or Cloud, the machine, or AI, will learn to articulate what is known and not known. At first it may need to engage us in a dialog to clarify ambiguities (as we humans do when answering questions), but, unlike us, the answer machine will not hesitate to provide deep, obscure, complex factual knowledge on any subject—if it exists.

But the chief consequence of reliable instant answers is *not* a harmony of satisfaction. Abundant answers simply generate more questions! In my experience, the easier it is to ask a question and the more useful the reply, the more questions I have. While the answer machine can expand answers infinitely, our time to form the next question is very limited. There is an asymmetry in the work needed to generate a good question versus the work needed to absorb an answer. Answers become cheap and questions become valuable—the inverse of the situation now. Pablo Picasso brilliantly anticipated this inversion in 1964 when he told the writer William Fifield, "Computers are useless. They only give you answers."

So at the end of the day, a world of supersmart ubiquitous answers encourages a quest for the perfect question. What makes a perfect question? Ironically, the best questions are not questions that lead to answers, because answers are on their way to becoming cheap and plentiful. A good question is worth a million good answers.

A good question is like the one Albert Einstein asked himself as a small boy—"What would you see if you were traveling on a beam of light?" That question launched the theory of relativity, $E=MC^2$, and the atomic age.

A good question is not concerned with a correct answer.

A good question cannot be answered immediately.

A good question challenges existing answers.

A good question is one you badly want answered once you hear it, but had no inkling you cared before it was asked.

A good question creates new territory of thinking.

A good question reframes its own answers.

A good question is the seed of innovation in science, technology, art, politics, and business.

A good question is a probe, a what-if scenario.

A good question skirts on the edge of what is known and not known, neither silly nor obvious.

A good question cannot be predicted.

A good question will be the sign of an educated mind.

A good question is one that generates many other good questions.

A good question may be the last job a machine will learn to do.

A good question is what humans are for.

————

What is it that we are making with our question-and-answer machine?

Our society is moving away from the rigid order of hierarchy toward the fluidity of decentralization. It is moving from nouns to verbs, from tangible products to intangible becomings. From fixed media to messy remixed media. From stores to flows. And the value engine is moving from the certainties of answers to the uncertainties of questions. Facts, order, and answers will always be needed and useful. They are not going away, and in fact, like microbial life and concrete materials, facts will continue to underpin the bulk of our civilization. But the most precious aspects, the most dynamic, most valuable, and most productive facets of our lives and new technology will lie in the frontiers, in the edges where uncertainty, chaos, fluidity, and questions dwell. The technologies of generating answers will continue to be essential, so much that answers will become omnipresent, instant, reliable, and just about free. But the technologies that help generate questions will be valued more. Question makers will be seen, properly, as the engines that generate the new fields, new industries, new brands, new possibilities, new continents that our restless species can explore. Questioning is simply more powerful than answering.

12

BEGINNING

Thousands of years from now, when historians review the past, our ancient time here at the beginning of the third millennium will be seen as an amazing moment. This is the time when inhabitants of this planet first linked themselves together into one very large thing. Later the very large thing would become even larger, but you and I are alive at that moment when it first awoke. Future people will envy us, wishing they could have witnessed the birth we saw. It was in these years that humans began animating inert objects with tiny bits of intelligence, weaving them into a cloud of machine intelligences and then linking billions of their own minds into this single supermind. This convergence will be recognized as the largest, most complex, and most surprising event on the planet up until this time. Braiding nerves out of glass, copper, and airy radio waves, our species began wiring up all regions, all processes, all people, all artifacts, all sensors, all facts and notions into a grand network of hitherto unimagined complexity. From this embryonic net was born a collaborative interface for our civilization, a sensing, cognitive apparatus with power that exceeded any previous invention. This megainvention, this organism, this machine—if we want to call it that—subsumes all the other machines made, so that in effect there is only one thing that permeates our lives to such a degree that it becomes essential to our identity. This very large thing provides a new way of thinking (perfect search, total recall, planetary scope) and a new mind for an old species. It is the Beginning.

The Beginning is a century-long process, and its muddling forward is mundane. Its big databases and extensive communications are boring. Aspects of this dawning real-time global mind are either dismissed as nonsense or feared. There is indeed a lot to be legitimately worried about because there is not a single aspect of human culture—or nature—that is left untouched by this syncopated pulse. Yet because we are the parts of something that has begun operating at a level above us, the outline of this emerging very large thing is obscured. All we know is that from its very beginning, it is upsetting the old order. Fierce pushback is to be expected.

What to call this very large masterpiece? Is it more alive than machine? At its core 7 billion humans, soon to be 9 billion, are quickly cloaking themselves with an always-on layer of connectivity that comes close to directly linking their brains to each other. A hundred years ago H. G. Wells imagined this large thing as the world brain. Teilhard de Chardin named it the noosphere, the sphere of thought. Some call it a global mind, others liken it to a global superorganism since it includes billions of manufactured silicon neurons. For simple convenience and to keep it short, I'm calling this planetary layer the holos. By holos I include the collective intelligence of all humans *combined with* the collective behavior of all machines, plus the intelligence of nature, plus whatever behavior emerges from this whole. This whole equals holos.

The scale of what we are becoming is simply hard to absorb. It is the largest thing we have made. Let's take just the hardware, for example. Today there are 4 billion mobile phones and 2 billion computers linked together into a seamless cortex around the globe. Add to them all the billions of peripheral chips and affiliated devices from cameras to cars to satellites. Already in 2015 a grand total of 15 billion devices have been wired up into one large circuit. Each of these devices contains 1 billion to 4 billion transistors themselves, so in total the holos operates with a sextillion transistors (10 with 21 zeros). These transistors can be thought of as the neurons in a vast brain. The human brain has roughly 86 billion neurons, or a trillion times fewer than the holos. In terms of magnitude, the holos already significantly exceeds our brains in complexity. And our brains are not doubling in size every few years. The holos mind is.

Today, the hardware of the holos acts like a very large virtual computer made up of as many computer chips as there are transistors in a computer. This virtual computer's top-level functions operate at approximately the speed of an early PC. It processes 1 million emails each second, and 1 million messages per second, which essentially means the holos currently runs at 1 megahertz. Its total external storage is about 600 exabytes today. In any one second, 10 terabits course through its backbone nerves. It has a robust immune system, weeding spam from its trunk lines and rerouting around damage as a type of self-healing.

And who will write the code that makes this global system useful and productive? We will. We think we are merely wasting time when we surf mindlessly or post an item for our friends, but each time we click a link we strengthen a node somewhere in the holos mind, thereby programming it by using it. Think of the 100 billion times *per day* humans click on a web page as a way of teaching the holos what we think is important. Each time we forge a link between words, we teach this contraption an idea.

This is the new platform that our lives will run on. International in scope. Always on. At current rates of technological adoption I estimate that by the year 2025 every person alive—that is, 100 percent of the planet's inhabitants—will have access to this platform via some almost-free device. Everyone will be on it. Or in it. Or, simply, everyone will be it.

This big global system will not be utopia. Even three decades from now, regional fences will remain in this cloud. Parts will be firewalled, censored, privatized. Corporate monopolies will control aspects of the infrastructure, though these internet monopolies are fragile and ephemeral, subject to sudden displacement by competitors. Although minimal access will be universal, higher bandwidth will be uneven and clumped around urban areas. The rich will get the premium access. In short, the distribution of resources will resemble the rest of life. But this is critical and transformative, and even the least of us will be part of it.

Right now, in this Beginning, this imperfect mesh spans 51 billion hectares, touches 15 billion machines, engages 4 billion human minds in real time, consumes 5 percent of the planet's electricity, runs at inhuman speeds, tracks half our daytime hours, and is the conduit for the majority flow of our money. The level of organization is a step above the largest things we have made till now: cities. This jump in levels reminds some physicists of a phase transition, the discontinuous break between a molecule's state—say, between ice and water, or water and steam. The difference in temperature or pressure separating two phases is almost trivial, but the fundamental reorganization across the threshold makes the material behave in a whole new manner. Water is definitely a different state than ice.

The large-scale, ubiquitous interconnection of this new platform at first seems like just the natural extension of our traditional society. It seems to just add digital relationships to our existing face-to-face relationships. We add a few more friends. We expand our network of acquaintances. Broaden our sources of news. Digitize our movements. But, in fact, as all these qualities keep steadily increasing, just as temperature and pressure slowly creep higher, we pass an inflection point, a complexity threshold, where the change is discontinuous—a phase transition—and suddenly we are in a new state: a different world with new normals.

We are in the Beginning of that process, right at the cusp of that discontinuity. In this new regime, old cultural forces, such as centralized authority and uniformity, diminish while new cultural forces, such as the ones I describe in this book—sharing, accessing, tracking—come to dominate our institutions and personal lives. As the new phase congeals, these forces will continue to intensify. Sharing, though excessive to some now, is just beginning. The switch from ownership to access has barely begun. Flows and streams are still trickles. While it seems as if we are tracked too much already, we'll be tracking a thousand times as much in the coming decades. Each one of these functions will be accelerated by high-quality cognification, just now being born, making the smartest things we do today seem very dumb. None of this is final. These transitions are but the first step in a process, a process of becoming. It is a Beginning.

———————

Look at a satellite photograph of the earth at night to get a glimpse of this very large organism. Brilliant clusters of throbbing city lights trace out organic patterns on the dark land. The cities gradually dim at their edges to form thin long lighted highways connecting other distant city clusters. The routes of lights outward are dendritic, treelike patterns. The image is deeply familiar. The cities are ganglions of nerve cells; the lighted highways are the axons of nerves, reaching to a synaptic connection. Cities are the neurons of the holos. We live inside this thing.

This embryonic very large thing has been running continuously for at least 30 years. I am aware of no other machine—of any type—that has run that long with zero downtime. While portions of it will probably spin down temporarily one day due to power outages or cascading infections, the entire thing is unlikely to go quiet in the coming decades. It has been and will likely remain the most reliable artifact we have.

This picture of an emerging superorganism reminds some scientists of the concept of "the singularity." A "singularity" is a term borrowed from physics to describe a frontier beyond which nothing can be known.

There are two versions in pop culture: a hard singularity and a soft singularity. The hard version is a future brought about by the triumph of a superintelligence. When we create an AI that is capable of making an intelligence smarter than itself, it can in theory make generations of ever smarter AIs. In effect, AI would bootstrap itself in an infinite accelerating cascade so that each smarter generation is completed faster than the previous generation until AIs very suddenly get so smart that they solve all existing problems in godlike wisdom and leave us humans behind. It is called a singularity because it is beyond what we can perceive. Some call that our "last invention." For various reasons, I think that scenario is unlikely.

A soft singularity is more likely. In this future scenario AIs don't get so smart that they enslave us (like evil versions of smart humans); rather AI and robots and filtering and tracking and all the technologies I outline in this book converge—humans plus machines—and together we move to a complex interdependence. At this level many phenomenon occur at scales greater than our current lives, and greater than we can perceive— which is the mark of a singularity. It's a new regime wherein our creations makes us better humans, but also one where we can't live without what we've made. If we have been living in rigid ice, this is liquid—a new phase state.

This phase change has already begun. We are marching inexorably toward firmly connecting all humans and all machines into a global matrix. This matrix is not an artifact, but a process. Our new supernetwork is a standing wave of change that steadily spills forward new arrangements of our needs and desires. The particular products, brands, and companies that will surround us in 30 years are entirely unpredictable. The specifics at that time hinge on the crosswinds of individual chance and fortune. But the overall direction of this large-scale vibrant process is clear and unmistakable. In the next 30 years the holos will continue to lean in the same direction it has for the last 30 years: toward increased flowing, sharing, tracking, accessing, interacting, screening, remixing, filtering, cognifying, questioning, and becoming. We stand at this moment at the Beginning.

The Beginning, of course, is just beginning.

ACKNOWLEDGMENTS

I am indebted to Paul Slovak, my editor at Viking, who has long supported my efforts to make sense of technology, and to my agent John Brockman, who suggested this book. For editorial guidance on the first draft I relied on Jay Schaefer, master book coach based in San Francisco. Librarian Camille Hartsell did most of the factual research and provided the extensive endnotes. Claudia Lamar assisted in research, fact-checking, and formatting help. Two of my former colleagues at *Wired*, Russ Mitchell and Gary Wolf, waded through an early rough draft and made important suggestions that I incorporated. Over the span of years that I wrote this material I benefited from the precious time of many interviewees. Among them were John Battelle, Michael Naimark, Jaron Lanier, Gary Wolf, Rodney Brooks, Brewster Kahle, Alan Greene, Hal Varian, George Dyson, and Ethan Zuckerman. Thanks to the editors of *Wired* and *The New York Times Magazine*, who were instrumental in shaping initial versions of portions of this book.

Most important, this book is dedicated to my family—Giamin, Kaileen, Ting, and Tywen—who keep me grounded and pointed forward. Thank you.

NOTES

1: BECOMING

11 **average lifespan of a phone app:** Erick Schonfeld, "Pinch Media Data Shows the Average Shelf Life of an iPhone App Is Less Than 30 Days," *TechCrunch*, February 19, 2009.

13 **sea pirates two centuries ago:** Peter T. Leeson, *The Invisible Hook: The Hidden Economics of Pirates* (Princeton, NJ: Princeton University Press, 2011).

15 **graphic Netscape browser:** Jim Clark and Owen Edwards, *Netscape Time: The Making of the Billion-Dollar Start-Up That Took on Microsoft* (New York: St. Martin's, 1999).

15 **not designed for doing commerce:** Philip Elmer-Dewitt, "Battle for the Soul of the Internet," *Time*, July 25, 1994.

15 **"The Internet? Bah!":** Clifford Stoll, "Why the Web Won't Be Nirvana," *Newsweek*, February 27, 1995 (original title: "The Internet? Bah!").

16 **"CB radio of the '90s":** William Webb, "The Internet: CB Radio of the 90s?," *Editor & Publisher*, July 8, 1995.

18 **Bush outlined the web's core idea:** Vannevar Bush, "As We May Think," *Atlantic*, July 1945.

18 **Nelson, who envisioned his own scheme:** Theodor H. Nelson, "Complex Information Processing: A File Structure for the Complex, the Changing and the Indeterminate," in *ACM '65: Proceedings of the 1965 20th National Conference* (New York: ACM, 1965), 84–100.

19 **"transclusion":** Theodor H. Nelson, *Literary Machines* (South Bend, IN: Mindful Press, 1980).

19 **"intertwingularity":** Theodor H. Nelson, *Computer Lib: You Can and Must Understand Computers Now* (South Bend, IN: Nelson, 1974).

20 **total number of web pages:** "How Search Works," Inside Search, Google, 2013, accessed April 26, 2015.

21 **90 billion searches a month:** Steven Levy, "How Google Search Dealt with Mobile," *Medium*, Backchannel, January 15, 2015.

22 **50 million blogs in the early 2000s:** David Sifry, "State of the Blogosphere, August 2006," Sifry's Alerts, August 7, 2006.

22 **65,000 per day are posted:** "YouTube Serves Up 100 Million Videos a Day Online," Reuters, July 16, 2006.

22 **300 video hours every minute, in 2015:** "Statistics," YouTube, April 2015, https://goo.gl/RVb7oz.

23 **women online first outnumbered men:** Deborah Fallows, "How Women and Men Use the Internet: Part 2—Demographics," Pew Research Center, December 28, 2005.

23 **51 percent of netizens are female:** Calculation based on "Internet User Demographics: Internet Users in 2014," Pew Research Center, 2014; and "2013 Population Estimates," U.S. Census Bureau, 2015.

23 **bone-creaking 44 years old:** Weighted average of internet users in 2014 based on "Internet User Demographics," Pew Research Center, 2014; and "2014 Population Estimates," U.S. Census Bureau, 2014.

25 **mcdonalds.com was still unclaimed:** Joshua Quittner, "Billions Registered," *Wired* 2(10), October 1994.

2: COGNIFYING

31 **several hundred "instances" of the AI:** Personal visit to IBM Research, June 2014.

31 **"world's best diagnostician":** Personal correspondence with Alan Greene.

32 **$18 billion in investments since 2009:** Private analysis by Quid, Inc., 2014.

32 **in-house AI research teams:** Reed Albergotti, "Zuckerberg, Musk Invest in Artificial-Intelligence Company," *Wall Street Journal*, March 21, 2014.

32 **purchased AI companies since 2014:** Derrick Harris, "Pinterest, Yahoo, Dropbox and the (Kind of) Quiet Content-as-Data Revolution," *Gigaom*, January 6, 2014; Derrick Harris "Twitter Acquires Deep Learning Startup Madbits," *Gigaom*, July 29, 2014; Ingrid Lunden, "Intel Has Acquired Natural Language Processing Startup Indisys, Price 'North' of $26M, to Build Its AI Muscle," *TechCrunch*, September 13, 2013; and Cooper Smith, "Social Networks Are Investing Big in Artificial Intelligence," *Business Insider*, March 17, 2014.

32 **expanding 70 percent a year:** Private analysis by Quid, Inc., 2014.

32 **taught an AI to learn to play:** Volodymyr Mnih, Koray Kavukcuoglu, David Silver, et al., "Human-Level Control Through Deep Reinforcement Learning," *Nature* 518, no. 7540 (2015): 529–33.

35 **Betterment or Wealthfront:** Rob Berger, "7 Robo Advisors That Make Investing Effortless," *Forbes*, February 5, 2015.

37 **80 percent of its revenue:** Rick Summer, "By Providing Products That Consumers Use Across the Internet, Google Can Dominate the Ad Market," Morningstar, July 17, 2015.

37 **3 billion queries that Google conducts:** Danny Sullivan, "Google Still Doing at Least 1 Trillion Searches Per Year," Search Engine Land, January 16, 2015.

37 **Google CEO Sundar Pichai stated:** James Niccolai, "Google Reports Strong Profit, Says It's 'Rethinking Everything' Around Machine Learning," *ITworld*, October 22, 2015.

37 **the AI winter:** "AI Winter," Wikipedia, accessed July 24, 2015.

38 **Billions of neurons in our brain:** Frederico A. C. Azevedo, Ludmila R. B. Carvalho, Lea T. Grinberg, et al., "Equal Numbers of Neuronal and Non-Neuronal Cells Make the Human Brain an Isometrically Scaled-up Primate Brain," *Journal of Comparative Neurology* 513, no. 5 (2009): 532–41.

38 **run neural networks in parallel:** Rajat Raina, Anand Madhavan, and Andrew Y. Ng, "Large-Scale Deep Unsupervised Learning Using Graphics Processors," *Proceedings of the 26th Annual International Conference on Machine Learning, ICML '09* (New York: ACM, 2009), 873–80.

39 **neural nets running on GPUs:** Klint Finley, "Netflix Is Building an Artificial Brain Using Amazon's Cloud," *Wired*, February 13, 2014.

39 **dozen examples as a child before it can distinguish:** Personal correspondence with Paul Quinn, Department of Psychological and Brain Sciences, University of Delaware, August 6, 2014.

39 **thousand games of chess:** Personal correspondence with Daylen Yang (author of the Stockfish chess app), Stefan Meyer-Kahlen (developed the multiple award-winning computer chess program Shredder), and Danny Kopec (American chess International Master and cocreator of one of the standard computer chess testing systems), September 2014.

39 **"akin to building a rocket ship":** Caleb Garling, "Andrew Ng: Why 'Deep Learning' Is a Mandate for Humans, Not Just Machines," *Wired*, May 5, 2015.

40 **In 2006, Geoff Hinton:** Kate Allen, "How a Toronto Professor's Research Revolutionized Artificial Intelligence," *Toronto Star*, April 17, 2015.

40 **he dubbed "deep learning":** Yann LeCun, Yoshua Bengio, and Geoffrey Hinton, "Deep Learning," *Nature* 521, no. 7553 (2015): 436–44.

40 **the network effect:** Carl Shapiro and Hal R. Varian, *Information Rules: A Strategic Guide to the Network Economy* (Boston: Harvard Business Review Press, 1998).

41 **famous man-versus-machine match:** "Deep Blue," IBM 100: Icons of Progress, March 7, 2012.

41 **rather than competes against them:** Owen Williams, "Garry Kasparov—Biography," KasparovAgent.com, 2010.

41 **freestyle chess matches:** Arno Nickel, Freestyle Chess, 2010.

41 **centaurs won 53 games:** Arno Nickel, "The Freestyle Battle 2014," Infinity Chess, 2015.

41 **several different chess programs:** Arno Nickel, "'Intagrand' Wins the Freestyle Battle 2014," Infinity Chess, 2015.

42 **grand master rating of all time:** "FIDE Chess Profile (Carlsen, Magnus)," World Chess Federation, 2015.

43 **AI that can view a photo portrait of any person:** Personal interview at Facebook, September 2014.

49 **70 percent of American workers:** U.S. Census Bureau, "Current Population Reports: Farm Population," *Persons in Farm Occupations: 1820 to 1987* (Washington, D.C.: U.S. Government Printing Office, 1988), 4.

49 **all but 1 percent of their jobs:** "Employed Persons by Occupation, Sex, and Age," Employment & Earnings Online, U.S. Bureau of Labor Statistics, 2015.

50 **this is a big deal:** Scott Santens, "Self-Driving Trucks Are Going to Hit Us Like a Human-Driven Truck," *Huffington Post*, May 18, 2015.

51 **accurate caption for any photo:** Tom Simonite, "Google Creates Software That Tells You What It Sees in Images," *MIT Technology Review*, November 18, 2014.

52 **Industrial robots cost $100,000-plus:** Angelo Young, "Industrial Robots Could Be 16% Less Costly to Employ Than People by 2025," *International Business Times*, February 11, 2015.

52 **four times that amount over a lifespan:** Martin Haegele, Thomas Skordas, Stefan Sagert, et al., "Industrial Robot Automation," White Paper FP6-001917, European Robotics Research Network, 2005.

53 **Priced at $25,000:** Angelo Young, "Industrial Robots Could Be 16% Less Costly to Employ Than People by 2025," *International Business Times*, February 11, 2015.

54 **all but seven minutes of a typical flight:** John Markoff, "Planes Without Pilots," *New York Times*, April 6, 2015.

3: FLOWING

62 **steady flow of household replenishables:** "List of Online Grocers," Wikipedia, accessed August 18, 2015.

63 **new medium imitates the medium it replaces:** Marshall McLuhan, *Culture Is Our Business* (New York: McGraw-Hill, 1970).

66 **top ten music videos:** "List of Most Viewed YouTube Videos," Wikipedia, accessed August 18, 2015.

72 **about $2.26 per download:** "Did Radiohead's 'In Rainbows' Honesty Box Actually Damage the Music Industry?," *NME*, October 15, 2012.

74 **create a chorus from it:** Eric Whitacre's Virtual Choir, "Lux Aurumque," March 21, 2010.

74 **containing 30 million tracks of music:** "Information," Spotify, accessed June 18, 2015.

75 **its 250 million fans:** Romain Dillet, "SoundCloud Now Reaches 250 Million Visitors in Its Quest to Become the Audio Platform of the Web," *TechCrunch*, October 29, 2013.

75 **27 percent of music sales:** Joshua P. Friedlander, "News and Notes on 2014 RIAA Music Industry Shipment and Revenue Statistics," Recording Industry Association of America, 2015, http://goo.gl/Ozgk8f.

75 **Spotify pays 70 percent:** "Spotify Explained," Spotify Artists, 2015.

75 **streaming takeover "is inevitable":** Joan E. Solsman, "Attention, Artists: Streaming Music Is the Inescapable Future. Embrace It," *CNET*, November 14, 2014.

76 **hours of music required:** Personal estimation.

76 **new podcasts launch every day:** Personal correspondence with Todd Pringle, GM and VP of Product, Stitcher, April 26, 2015.

78 **four ways books embody fixity:** Nicholas Carr, "Words in Stone and on the Wind," Rough Type, February 3, 2012.

4: SCREENING

85 **50,000 words in Old English to a million:** Robert McCrum, Robert MacNeil, and William Cran, *The Story of English*, third revised ed. (New York: Penguin Books, 2002); and *Encyclopedia Americana*, vol. 10 (Grolier, 1999).

86 **romance novel was invented in 1740:** Pamela Regis, *A Natural History of the Romance Novel* (Philadelphia: University of Pennsylvania Press, 2007).

86 **three quarters of the towns:** Calculation based on approximately 1,700 public libraries and 2,269 places with a population of 2,500 or higher. Florence Anderson, *Carnegie Corporation Library Program 1911–1961* (New York: Carnegie Corporation, 1963); Durand R. Miller, *Carnegie Grants for Library Buildings, 1890–1917* (New York: Carnegie Corporation, 1943); and "1990 Census of Population and Housing," U.S. Census Bureau, CPH21, 1990.

86 **5 billion digital screens illuminate our lives:** Extrapolation based on "Installed Base of Internet-Connected Video Devices to Exceed Global Population in 2017," IHS, October 8, 2013.

86 **3.8 billion new additional screens per year:** 2014 Total Global Shipments, IHS Display Search; personal communication with Lee Graham, May 1, 2015.

89 **reading scores trended down:** "Average SAT Scores of College-Bound Seniors," College Board, 2015, http://goo.gl/Rbmu0q.

89 **tripled since 1980:** Roger E. Bohn and James E. Short, *How Much Information? 2009 Report on American Consumers*, Global Information Industry Center, University of California, San Diego, 2009.

89 **60 trillion pages:** "How Search Works," Inside Search, Google, 2013.

89 **80 million blog posts per day:** Sum of 2 million on WordPress, 78 million on Tumblr: "A Live Look at Activity Across WordPress.com," WordPress, April 2015; and "About (Posts Today)," Tumblr, accessed August 5, 2015.

89 **500 million quips per day:** "About (Tweets Sent Per Day)," Twitter, August 5, 2015.

91 **Some scholars of literature:** Sven Birkerts, "Reading in a Digital Age," *American Scholar*, March 1, 2010.

91 **Neurological studies show:** Stanislas Dehaene, *Reading in the Brain: The Science and Evolution of a Human Invention* (New York: Viking, 2009).

91 **screen only one word wide:** "Rapid Serial Visual Presentation," Wikipedia, accessed June 24, 2015.

92 **36 million Kindles and ebook readers:** Helen Ku, "E-Ink Forecasts Loss as Ebook Device Demand Falls," *Taipei Times*, March 29, 2014.

92–93 **books that are projected wide and big:** Stefan Marti, "TinyProjector," MIT Media Lab, October 2000–May 2002.

96 **concepts elsewhere in the encyclopedia:** "List of Wikipedias," Wikimedia Meta-Wiki, accessed April 30, 2015.

96 **great library at Alexandria:** Lionel Casson, *Libraries in the Ancient World* (New Haven, CT: Yale University Press, 2001); Andrew Erskine, "Culture and Power in Ptolemaic Egypt: The Library and Museum at Alexandria," *Greece and Rome* 42 (1995).

96 **backing up the entire internet:** Personal correspondence with Brewster Kahle, 2006.

97 **at least 310 million books:** "WorldCat Local," WorldCat, accessed August 18, 2015.

97 **1.4 billion articles and essays:** Ibid.

97 **180 million songs:** "Introducing Gracenote Rhythm," Gracenote, accessed May 1, 2015.

97 **3.5 trillion images:** "How Many Photos Have Ever Been Taken?," *1,000 Memories* blog, April 10, 2012, accessed via Internet Archive, May 2, 2015.

97 **330,000 movies:** "Database Statistics," IMDb, May 2015.

97 **1 billion hours of videos, TV shows, and short films:** Inferred from "Statistics," YouTube, accessed August 18, 2015.

97 **60 trillion public web pages:** "How Search Works," Inside Search, Google, 2013.

97 **50-petabyte hard disks:** Private communication with Brewster Kahle, 2006.

102 **25 million orphan works:** Naomi Korn, *In from the Cold: An Assessment of the Scope of 'Orphan Works' and Its Impact on the Delivery of Services to the Public,* JISC Content, Collections Trust, Cambridge, UK, April 2009.

103 **"stories, not of atoms":** Muriel Rukeyser, *The Speed of Darkness: Poems* (New York: Random House, 1968).

103 **what we are paying attention to:** Phillip Moore, "Eye Tracking: Where It's Been and Where It's Going," User Testing, June 4, 2015.

103 **read our emotions as we read the screen:** Mariusz Szwoch and Wioleta Szwoch, "Emotion Recognition for Affect Aware Video Games," in *Image Processing & Communications Challenges* 6, ed. Ryszard S. Choraś, *Advances in Intelligent Systems and Computing* 313, Springer International, 2015, 227–36.

105 **informational layer to reality:** Jessi Hempel, "Project Hololens: Our Exclusive Hands-On with Microsoft's Holographic Goggles," *Wired*, January 21, 2015; and Sean Hollister, "How Magic Leap Is Secretly Creating a New Alternate Reality," *Gizmodo*, November 9, 2014.

5: ACCESSING

109 ***TechCrunch* recently observed:** Tom Goodwin, "The Battle Is for the Customer Interface," *TechCrunch*, March 3, 2015.

109 **800,000-volume library:** "Kindle Unlimited," Amazon, accessed June 24, 2015.

110 **tin-coated steel and it weighed:** Chaz Miller, "Steel Cans," Waste 360, March 1, 2008.

110 **one fifth of its original weight:** "Study Finds Aluminum Cans the Sustainable Package of Choice," Can Manufacturers Institute, May 20, 2015.

110 **weight of the average automobile has fallen:** Ronald Bailey, "Dematerializing the Economy," Reason.com, September 5, 2001.

111 **In 1930 it took only one kilogram:** Sylvia Gierlinger and Fridolin Krausmann, "The Physical Economy of the United States of America," *Journal of Industrial Ecology* 16, no. 3 (2012): 365–77, Figure 4a.

111 **from $1.64 in 1977 to $3.58 in 2000:** Figures adjusted for inflation. Ronald Bailey, "Dematerializing the Economy," Reason.com, September 5, 2001.

111 **"Software eats everything":** Marc Andreessen, "Why Software Is Eating the World," *Wall Street Journal*, August 20, 2011.

113 **Toffler called in 1980 the "prosumer":** Alvin Toffler, *The Third Wave* (New York: Bantam, 1984).

113 **subscribe to Photoshop:** "Subscription Products Boost Adobe Fiscal 2Q Results," Associated Press, June 16, 2015.

115 **Uber for laundry:** Jessica Pressler, "'Let's, Like, Demolish Laundry,'" *New York*, May 21, 2014.

115 **Uber for doctor house calls:** Jennifer Jolly, "An Uber for Doctor House Calls," *New York Times*, May 5, 2015.

117 **sizable bag rental business:** Emily Hamlin Smith, "Where to Rent Designer Handbags, Clothes, Accessories and More," *Cleveland Plain Dealer*, September 12, 2012.

120 **phone app, such as M-Pesa:** Murithi Mutiga, "Kenya's Banking Revolution Lights a Fire," *New York Times*, January 20, 2014.

120 **has $3 billion in circulation:** "Bitcoin Network," Bitcoin Charts, accessed June 24, 2015.

120 **100,000 vendors accepting the coins:** Wouter Vonk, "Bitcoin and BitPay in 2014," *BitPay* blog, February 4, 2015.

121 **Six times an hour:** Colin Dean, "How Many Bitcoin Are Mined Per Day?," Bitcoin Stack Exchange, March 28, 2013.

127 **Knowledge-Based Trust:** Hal Hodson, "Google Wants to Rank Websites Based on Facts Not Links," *New Scientist*, February 28, 2015.

127 **tools are extensions of our selves:** Marshall McLuhan, *Understanding Media: The Extensions of Man* (New York: McGraw-Hill, 1964).

128 **down only 14 minutes in 2014:** Brandon Butler, "Which Cloud Providers Had the Best Uptime Last Year?," *Network World*, January 12, 2015.

129 **app onto their phones called FireChat:** Noam Cohen, "Hong Kong Protests Propel FireChat Phone-to-Phone App," *New York Times*, October 5, 2014.

6: SHARING

135 **"new modern-day sort of communists":** Michael Kanellos, "Gates Taking a Seat in Your Den," CNET, January 5, 2005.

135 **first collaborative web page in 1994:** Ward Cunningham, "Wiki History," March 25, 1995, http://goo.gl/2qAjTO.

135–36 **tracks nearly 150 wiki engines today:** "Wiki Engines," accessed June 24, 2015, http://goo.gl/5auMv6.

136 **billion instances of Creative Commons:** "State of the Commons," Creative Commons, accessed May 2, 2015.

138 **"dot-communism":** Theta Pavis, "The Rise of Dot-Communism," *Wired*, October 25, 1999.

138 **"composed entirely of free agents":** Roshni Jayakar, "Interview: John Perry Barlow, Founder of the Electronic Frontier Foundation," *Business Today*, December 6, 2000, accessed July 30, 2015, via Internet Archive, April 24, 2006.

138 **ranked by the increasing degree of coordination:** Clay Shirky, *Here Comes Everybody: The Power of Organizing Without Organizations* (New York: Penguin Press, 2008).

139 **1.8 billion per day:** Mary Meeker, "Internet Trends 2014—Code Conference," Kleiner Perkins Caufield & Byers, 2014.

139 **billions of videos served by YouTube:** "Statistics," YouTube, accessed June 24, 2015.

139 **millions of fan-created stories:** Piotr Kowalczyk, "15 Most Popular Fanfiction Websites," Ebook Friendly, January 13, 2015.

140 **the socialist promise:** "From Each According to His Ability, to Each According to His Need," Wikipedia, accessed June 24, 2015.

141 **Half of all web pages in the world today:** "July 2015 Web Server Survey," Netcraft, July 22, 2015.

141 **more than 35 million servers:** Jean S. Bozman and Randy Perry, "Server Transition Alternatives: A Business Value View Focusing on Operating Costs," White Paper 231528R1, IDC, 2012.

141 **running free Apache software:** "July 2015 Web Server Survey," Netcraft, July 22, 2015.

141 **3D Warehouse offers several million:** "Materialise Previews Upcoming Printables Feature for Trimble's 3D Warehouse," Materialise, April 24, 2015.

141 **community-designed Arduinos:** "Arduino FAQ—With David Cuartielles," Medea, April 5, 2013.

141 **Raspberry Pi computers:** "About 6 Million Raspberry Pis Have Been Sold," Adafruit, June 8, 2015.

142 **"alternative to both state-based":** Yochai Benkler, *The Wealth of Networks: How Social Production Transforms Markets and Freedom* (New Haven, CT: Yale University Press, 2006).

143 **650,000 people:** "Account Holders," Black Duck Open Hub, accessed June 25, 2015.

143 **more than half a million projects:** "Projects," Black Duck Open Hub, accessed June 25, 2015.

143 **size of the General Motors workforce:** "Annual Report 2014," General Motors, 2015, http://goo.gl/DhXIxp.

143 **several hundred contributors:** "Current Apache HTTP Server Project Members," Apache HTTP Server Project, accessed June 25, 2015.

143 **60,000 person-years of work:** Amanda McPherson, Brian Proffitt, and Ron Hale-Evans, "Estimating the Total Development Cost of a Linux Distribution," Linux Foundation, 2008.

143 **10,000 daily active communities:** "About Reddit," Reddit, accessed June 25, 2015.

143 **1 billion monthly users:** "Statistics," YouTube, accessed June 25, 2015.

143 **have contributed to Wikipedia:** "Wikipedia: Wikipedians," Wikipedia, accessed June 25, 2015.

143 **posted on Instagram:** "Stats," Instagram, accessed May 2, 2015.

143 **700 million groups participate in Facebook:** "Facebook Just Released Their Monthly Stats and the Numbers Are Staggering," TwistedSifter, April 23, 2015.

144 **1.4 billion citizens of Facebook:** Ibid.

144 **survey of 2,784 open source developers:** Rishab Aiyer Ghosh, Ruediger Glott, Bernhard Krieger, et al., "Free/Libre and Open Source Software: Survey and Study," International Institute of Infonomics, University of Maastricht, Netherlands, 2002, Figure 35: "Reasons to Join and to Stay in OS/FS Community."

144 **"improve my own damn software":** Gabriella Coleman, "The Political Agnosticism of Free and Open Source Software and the Inadvertent Politics of Contrast," *Anthropological Quarterly* 77, no. 3 (2004): 507–19.

145 **it had only 30 employees:** Gary Wolf, "Why Craigslist Is Such a Mess," *Wired* 17(9), August 24, 2009.

148 **"as smart as everyone":** Larry Keeley, "Ten Commandments for Success on the Net," *Fast Company*, June 30, 1996.

148 **as Clay Shirky puts it:** Clay Shirky, *Here Comes Everybody: The Power of Organizing Without Organizations* (New York: Penguin Press, 2008).

149 **"Declaration of the Independence of Cyberspace":** John Perry Barlow, "Declaring Independence," *Wired* 4(6), June 1996.

149 **$24 billion in 2015:** Steven Perlberg, "Social Media Ad Spending to Hit $24 Billion This Year," *Wall Street Journal*, April 15, 2015.

150 **tried to harness readers' reports:** Rachel McAthy, "Lessons from the *Guardian*'s Open Newslist Trial," Journalism.co.uk, July 9, 2012.

150 *OhMyNews* **in South Korea:** "OhMyNews," Wikipedia, accessed July 30, 2015.

150 *Fast Company* **signed up 2,000:** Ed Sussman, "Why Michael Wolff Is Wrong," *Observer*, March 20, 2014.

151 **smaller number of editors:** Aaron Swartz, "Who Writes Wikipedia?," Raw Thought, September 4, 2006.

151 **"an old-boy network":** Kapor first said this about the internet pre-web in the late 1980s. Personal communication.

152 **not exactly a bastion of equality:** "Wikipedia: WikiProject Countering Systemic Bias," Wikipedia, accessed July 31, 2015.

154 **9,000 startups in 2015:** Mesh, accessed August 18, 2015, http://meshing.it.

155 **Babylonian chants:** Stef Conner, "The Lyre Ensemble," StefConner.com, accessed July 31, 2015.

155 **TV commercial using smartphones:** Amy Keyishian and Dawn Chmielewski, "Apple Unveils TV Commercials Featuring Video Shot with iPhone 6," *Re/code*, June 1, 2015; and V. Renée, "This New Ad for Bentley Was Shot on the iPhone 5S and Edited on an iPad Air Right Inside the Car," No Film School, May 17, 2014.

155 **paintings using an iPad:** Claire Cain Miller, "IPad Is an Artist's Canvas for David Hockney," Bits Blog, *New York Times*, January 10, 2014.

155 **Korean pop dance video "Gangnam Style":** Officialpsy, "Psy—Gangnam Style M/V," YouTube, July 15, 2012, accessed August 19, 2015, https://goo.gl/LoetL.

156 **9 million fans to fund 88,000 projects:** "Stats," Kickstarter, accessed June 25, 2015.

156 **raise more than $34 billion each year:** "Global Crowdfunding Market to Reach $34.4B in 2015, Predicts Massolution's 2015 CF Industry Report," Crowdsourcing.org, April 7, 2015.

156 **about 20,000 people who raised:** "The Year in Kickstarter 2013," Kickstarter, January 9, 2014.

157 **unless the total amount is raised:** "Creator Handbook: Funding," Kickstarter, accessed July 31, 2015.

157 **highest grossing Kickstarter campaign:** Pebble Time is currently the most funded Kickstarter, with $20,338,986 to date. "Most Funded," Kickstarter, accessed August 18, 2015.

157 **40 percent of all projects succeed:** "Stats: Projects and Dollars Success Rate," Kickstarter, accessed July 31, 2015.

159 **SeedInvest and FundersClub:** Marianne Hudson, "Understanding Crowdfunding and Emerging Trends," *Forbes*, April 9, 2015.

159 **ordinary citizens in early 2016:** Steve Nicastro, "Regulation A+ Lets Small Businesses Woo More Investors," *NerdWallet Credit Card* blog, June 25, 2015.

159 **more than $725 million:** "About Us: Latest Statistics," Kiva, accessed June 25, 2015.

160 **loans worth more than $10 billion:** Simon Cunningham, "Default Rates at Lending Club & Prosper: When Loans Go Bad," LendingMemo, October 17, 2014; and Davey Alba, "Banks Are Betting Big on a Startup That Bypasses Banks," *Wired*, April 8, 2015.

160 **GE has launched over 400 new products:** Steve Lohr, "The Invention Mob, Brought to You by Quirky," *New York Times*, February 14, 2015.

160 **Netflix announced an award:** Preethi Dumpala, "Netflix Reveals Million-Dollar Contest Winner," *Business Insider*, September 21, 2009.

160 **Forty thousand groups submitted:** "Leaderboard," Netflix Prize, 2009.

161 **150,000 car fanatics:** Gary Gastelu, "Local Motors 3-D-Printed Car Could Lead an American Manufacturing Revolution," Fox News, July 3, 2014.

161 **3-D-printed electric car:** Paul A. Eisenstein, "Startup Plans to Begin Selling First 3-D-Printed Cars Next Year," NBC News, July 8, 2015.

7: FILTERING

165 **8 million new songs:** Private correspondence with Richard Gooch, CTO, International Federation of the Phonographic Industry, April 15, 2015. This is a low estimate, with a higher estimate being 12 million, according to Paul Jessop and David Hughes, "In the Matter of: Technological Upgrades to Registration and Recordation Functions," Docket No. 2013-2, U.S. Copyright Office, 2013, Comments in response to the March 22, 2013, Notice of Inquiry.

165 **2 million new books:** "Annual Report," International Publishers Association, Geneva, 2014, http://goo.gl/UNfZLP.

165 **16,000 new films:** "Most Popular TV Series/Feature Films Released in 2014 (Titles by Country)," IMDb, 2015, accessed August 5, 2015.

165 **30 billion blog posts:** Extrapolations based on the following: "About (Posts Today)," Tumblr, accessed August 5, 2015; and "A Live Look at Activity Across WordPress.com," WordPress, accessed August 5, 2015.

165 **182 billion tweets:** "Company," Twitter, accessed August 5, 2015.

165 **400,000 new products:** "Global New Products Database," Mintel, accessed June 25, 2015.

165 **total number of songs:** "Introducing Gracenote Rhythm," Gracenote, accessed May 1, 2015.

168 **2,000 hours to completely read:** Based on an average reading speed of 250 words per minute, average for U.S. eighth graders. Brett Nelson, "Do You Read Fast Enough to Be Successful?," *Forbes*, June 4, 2012.

168 **29 million words:** "Great Books of the Western World," *Encyclopaedia Britannica Australia*, 2015.

169 **a third of Amazon sales:** James Manyika, Michael Chui, Brad Brown, et al., "Big Data: The Next Frontier for Innovation, Competition, and Productivity," McKinsey Global Institute, 2011. This is a conservative estimate. An outside analyst estimates it could be closer to two thirds.

169 **about $30 billion in 2014:** Extrapolated from 2014 sales/revenue of $88.9 billion. "Amazon.com Inc. (Financials)," *Market Watch*, accessed August 5, 2015.

169 **300 people working:** Janko Roettgers, "Netflix Spends $150 Million on Content Recommendations Every Year," *Gigaom*, October 9, 2014.

170 **automatically map one's position:** Eduardo Graells-Garrido, Mounia Lalmas, and Daniele Quercia, "Data Portraits: Connecting People of Opposing Views," arXiv Preprint, November 19, 2013.

170 **Studies show that going to the next circle:** Eytan Bakshy, Itamar Rosenn, Cameron Marlow, et al., "The Role of Social Networks in Information Diffusion," arXiv, January 2012, 1201.4145 [physics].

171 **200 average friends:** Aaron Smith, "6 New Facts About Facebook," Pew Research Center, February 3, 2014.

171 **all the posts your friends make:** Victor Luckerson, "Here's How Your Facebook News Feed Actually Works," *Time*, July 9, 2015.

172 **35 billion emails a day:** My calculation based on figures from the following: "Email Statistics Report, 2014–2018," Radicati Group, April 2014; and "Email Client Market Share," Litmus, April, 2015.

172 **filters the content of 60 trillion pages:** "How Search Works," Inside Search, Google, 2013.

172 **about 2 million times every minute:** Danny Sullivan, "Google Still Doing at Least 1 Trillion Searches Per Year," Search Engine Land, January 16, 2015.

172 **"3 billion questions a day":** Ibid.

176 **"a poverty of attention":** Herbert Simon, "Designing Organizations for an Information-Rich World," in *Computers, Communication, and the Public Interest*, ed. Martin Greenberger (Baltimore: Johns Hopkins University Press, 1971).

176 **TV still captures most of our attention:** Dounia Turrill and Glenn Enoch, "The Total Audience Report: Q1 2015," Nielsen, June 23, 2015.

177 **average CPM of various media platforms:** "The Media Monthly," Peter J. Solomon Company, 2014.

177 **half a trillion hours:** Calculation based on the following: "Census Bureau Projects U.S. and World Populations on New Year's Day," U.S. Census Bureau Newsroom, December 29, 2014; and Dounia Turrill and Glenn Enoch, "The Total Audience Report: Q1 2015," Nielsen, June 23, 2015.

177 **average $3.60 per hour of attention:** Michael Johnston, "What Are Average CPM Rates in 2014?," MonetizePros, July 21, 2014.

178 **4.3 hours to read:** Calculation based on Gabe Habash, "The Average Book Has 64,500 Words," *Publishers Weekly*, March 6, 2012; and Brett Nelson, "Do You Read Fast Enough to Be Successful?" *Forbes*, June 4, 2012.

178 **$23 to buy:** Private communication with Kempton Mooney, Nielsen, April 16, 2015.

180 **every one of the 60 trillion pages:** "How Search Works," Inside Search, Google, 2013.

180 **guided by the context:** "How Ads Are Targeted to Your Site," AdSense Help, accessed August 6, 2015.

180 **interest of the reader visiting:** Jon Mitchell, "What Do Google Ads Know About You?," ReadWrite, November 10, 2011.

182 **21 percent of Google's total revenue:** "2014 Financial Tables," Google Investor Relations, accessed August 7, 2015.

185 **5,000 user-made submissions:** Michael Castillo, "Doritos Reveals 10 'Crash the Super Bowl' Ad Finalists," *Adweek*, January 5, 2015.

185 **awards $1 million to the winner:** Gabe Rosenberg, "How Doritos Turned User-Generated Content into the Biggest Super Bowl Campaign of the Year," Content Strategist, Contently, January 12, 2015.

185 **4,000 were negative ads:** Greg Sandoval, "GM Slow to React to Nasty Ads," CNET, April 3, 2006.

186 **asymmetry of attention in email:** Esther Dyson, "Caveat Sender!," Project Syndicate, February 20, 2013.

187 **total lifetime spending of a customer:** Brad Sugars, "How to Calculate the Lifetime Value of a Customer," *Entrepreneur*, August 8, 2012.

187 **$168,000 worth of merchandise:** Morgan Quinn, "The 2015 Oscar Swag Bag Is Worth $168,000 but Comes with a Catch," *Las Vegas Review-Journal*, February 22, 2015.

189 **"downward trend in real commodity prices":** Paul Cashin and C. John McDermott, "The Long-Run Behavior of Commodity Prices: Small Trends and Big Variability," IMF Staff Papers 49, no. 2 (2002).

189 **dropping cost of copper:** Indur M. Goklany, "Have Increases in Population, Affluence and Technology Worsened Human and Environmental Well-Being?," *Electronic Journal of Sustainable Development* 1, no. 3 (2009).

190 **Luxury entertainment is increasing 6.5 percent:** Liyan Chen, "The Forbes 400 Shopping List: Living the 1% Life Is More Expensive Than Ever," *Forbes*, September 30, 2014.

190 **Spending at restaurants and bars:** Hiroko Tabuchi, "Stores Suffer from a Shift of Behavior in Buyers," *New York Times*, August 13, 2015.

190 **price of the average concert ticket:** Alan B. Krueger, "Land of Hope and Dreams: Rock and Roll, Economics, and Rebuilding the Middle Class," remarks given at the Rock and Roll Hall of Fame, White House Council of Economic Advisers, June 12, 2013.

190 **rose 400 percent from 1982 to 2014:** "Consumer Price Index for All Urban Consumers: Medical Care [CPIMEDSL]," U.S. Bureau of Labor Statistics, via FRED, Federal Reserve Bank of St. Louis, accessed June 25, 2015.

190 **rate for babysitting:** "2014 National Childcare Survey: Babysitting Rates & Nanny Pay," Urban Sitter, 2014; and Ed Halteman, "2013 INA Salary and Benefits Survey," International Nanny Association, 2012.

190 **cost of home visits:** Brant Morefield, Michael Plotzke, Anjana Patel, et al., "Hospice Cost Reports: Benchmarks and Trends, 2004–2011," Centers for Medicare and Medicaid Services, U.S. Department of Health and Human Services, 2011.

8: REMIXING

193 **existing resources that are rearranged:** Paul M. Romer, "Economic Growth," Concise Encyclopedia of Economics, Library of Economics and Liberty, 2008.

193 **combination of existing technologies:** W. Brian Arthur, *The Nature of Technology: What It Is and How It Evolves* (New York: Free Press, 2009).

194 **fan-created works to date:** Archive of Our Own, accessed July 29, 2015.

194 **12 million Vine clips:** Jenna Wortham, "Vine, Twitter's New Video Tool, Hits 13 Million Users," *Bits* blog, *New York Times*, June 3, 2013.

194 **1.5 billion daily loops:** Carmel DeAmicis, "Vine Rings in Its Second Year by Hitting 1.5 Billion Daily Loops," *Gigaom*, January 26, 2015.

196 **million person-hours to produce:** Personal calculation. Very few materials are consumed making a movie; 95 percent of the cost goes to labor and people's time, including subcontractors. Assuming that the average wage is less than $100 per hour, a $100 million movie entails at least one million hours of work.

196 **about 600 feature films are released:** "Theatrical Market Statistics 2014," Motion Picture Association of America, 2015.

197 **12 billion times in a single month:** "ComScore Releases January 2014 U.S. Online Video Rankings," comScore, February 21, 2014.

197 **more than any blockbuster movie:** The top-selling movie, *Gone with the Wind*, has sold an estimated 202,044,600 tickets. "All Time Box Office," Box Office Mojo, accessed August 7, 2015.

197 **100 million short video clips:** Mary Meeker, "Internet Trends 2014—Code Conference," Kleiner Perkins Caufield & Byers, 2014.

198 **Iron Editor challenge:** "Sakura-Con 2015 Results (and Info)," Iron Editor, April 7, 2015; and Neda Ulaby, "'Iron Editors' Test Anime Music-Video Skills," NPR, August 2, 2007.

198 **than with traditional cinematography:** Michael Rubin, *Droidmaker: George Lucas and the Digital Revolution* (Gainesville, FL: Triad Publishing, 2005).

199 **1.5 trillion photos posted:** Mary Meeker, "Internet Trends 2014—Code Conference," Kleiner Perkins Caufield & Byers, 2014.

200 **"database cinema":** Lev Manovich, "Database as a Symbolic Form," *Millennium Film Journal* 34 (1999); and Cristiano Poian, "Investigating Film Algorithm: Transtextuality in the Age of Database Cinema," presented at the Cinema and Contemporary Visual Arts II, V Magis Gradisca International Film Studies Spring School, 2015, accessed August 19, 2015.

201 **in the 13th century:** Malcolm B. Parkes, "The Influence of the Concepts of Ordinatio and Compilatio on the Development of the Book," in *Medieval Learning and Literature: Essays Presented to Richard William Hunt*, eds. J. J. G. Alexander and M. T. Gibson (Oxford: Clarendon Press, 1976), 115–27.

201 **Footnotes, invented in about:** Ivan Illich, *In the Vineyard of the Text: A Commentary to Hugh's Didascalicon* (Chicago: University of Chicago Press, 1996), 97.

201 **bibliographic citations:** Malcolm B. Parkes, "The Influence of the Concepts of Ordinatio and Compilation on the Development of the Book," in *Medieval Learning and Literature: Essays Presented to Richard William Hunt*, eds. J.J.G. Alexander and M. T. Gibson (Oxford: Clarendon Press, 1976), 115–27.

203 **gaining visual intelligence rapidly:** John Markoff, "Researchers Announce Advance in Image-Recognition Software," *New York Times*, November 17, 2014.

204 **"one can only reread it":** Vladimir Nabokov, *Lectures on Literature* (New York: Harcourt Brace Jovanovich, 1980).

208 **"He who receives an idea from me":** Thomas Jefferson, "Thomas Jefferson to Isaac McPherson, 13 Aug. 1813," in *Founders' Constitution*, eds. Philip B. Kurland and Ralph Lerner (Indianapolis: Liberty Fund, 1986).

208 **multibillion-dollar industry:** "Music Industry Revenue in the U.S. 2014," Statista, 2015, accessed August 11, 2015.

208 **uncertainty about Google's reuse:** Margaret Kane, "Google Pauses Library Project," CNET, October 10, 2005.

209 **70 years after the death of the creator:** "Duration of Copyright," Section 302(a), Circular 92, *Copyright Law of the United States of America and Related Laws Contained in Title 17 of the United States Code*, U.S. Copyright Office, accessed August 11, 2015.

9: INTERACTING

213 **dim room in the research labs of Stanford:** In-person VR demonstration by Jeremy Bailenson, director, Stanford University's Virtual Human Interaction Lab, June 2015.

215 **empty head-mounted display unit:** Menchie Mendoza, "Google Cardboard vs. Samsung Gear VR: Which Low-Cost VR Headset Is Best for Gaming?," *Tech Times*, July 21, 2015.

216 **"light field" projection:** Douglas Lanman, "Light Field Displays at AWE2014 (Video)," presented at the Augmented World Expo, June 2, 2014.

216 **first commercial light field units:** Jessi Hempel, "Project HoloLens: Our Exclusive Hands-On with Microsoft's Holographic Goggles," *Wired*, January 21, 2015.

218 **50,000 avatars are simultaneously roaming:** Luppicini Rocci, *Moral, Ethical, and Social Dilemmas in the Age of Technology: Theories and Practice* (Hershey, PA: IGI Global, 2013); and Mei Douthitt, "Why Did Second Life Fail? (Mei's Answer)," Quora, March 18, 2015.

218 **Half of them are there for virtual sex:** Frank Rose, "How Madison Avenue Is Wasting Millions on a Deserted Second Life," *Wired*, July 24, 2007.

219 **urinal in the men's restroom:** Nicholas Negroponte, "Sensor Deprived," *Wired* 2(10), October 1, 1994.

221 **"not enough Africa in them":** Kevin Kelly, "Gossip Is Philosophy," *Wired* 3(5), May 1995.

225 **Project Jacquard:** Virginial Postre, "Google's Project Jacquard Gets It Right," *BloombergView*, May 31, 2015.

225 **prototype from Northeastern University:** Brian Heater, "Northeastern University Squid Shirt Torso-On," Engadget, June 12, 2012.

225 **Sensory Substitution Vest:** Shirley Li, "The Wearable Device That Could Unlock a New Human Sense," *Atlantic*, April 14, 2015.

225 **she could drink from it:** Leigh R. Hochberg, Daniel Bacher, Beata Jarosiewicz, et al., "Reach and Grasp by People with Tetraplegia Using a Neurally Controlled Robotic Arm," *Nature* 485, no. 7398 (2012): 372–75.

228 **"skin him afterward":** Scott Sharkey, "Red Dead Redemption Review," 1Up.com, May 17, 2010.

228 **40 to 50 hours to complete:** "Red Dead Redemption," How Long to Beat, accessed August 11, 2015.

10: TRACKING

238 **200 Quantified Self Meetup groups:** "Quantified Self Meetups," Meetup, accessed August 11, 2015.

239 **he generates an annual report:** Nicholas Felton, "2013 Annual Report," Feltron.com, 2013.

243 **as if he could feel a map:** Sunny Bains, "Mixed Feelings," *Wired* 15(4), 2007.

245 **"calendar items, to-do lists":** Eric Thomas Freeman, "The Lifestreams Software Architecture" [dissertation], Yale University, May 1997.

245 **"Your entire cyberlife is right there":** Nicholas Carreiro, Scott Fertig, Eric Freeman, and David Gelernter, "Lifestreams: Bigger Than Elvis," Yale University, March 25, 1996.

247 **Steve Mann in the 1990s:** Steve Mann, personal web page, accessed July 29, 2015.

247 **Bell documented every aspect:** "MyLifeBits—Microsoft Research," Microsoft Research, accessed July 29, 2015.

252 **34 billion internet-enabled devices:** "The Internet of Things Will Drive Wireless Connected Devices to 40.9 Billion in 2020," ABI Research, August 20, 2014.

257 **600 percent increase in iPods:** "Apple's Profit Soars Thanks to iPod's Popularity," Associated Press, April 14, 2005.

257 **production tanked in 2009:** "Infographic: The Decline of iPod," Infogram, accessed May 3, 2015.

259 **benefits we humans covet:** Sean Madden, "Tech That Tracks Your Every Move Can Be Convenient, Not Creepy," *Wired*, March 10, 2014.

267 **54 billion sensors every year by 2020:** "Connections Counter: The Internet of Everything in Motion," The Network, Cisco, July 29, 2013.

11: QUESTIONING

270 **35 million articles in 288 languages:** "List of Wikipedias," Wikimedia Meta-Wiki, accessed April 30, 2015.

280 **"how to make people click ads":** Ashlee Vance, "This Tech Bubble Is Different," *Bloomberg Business*, April 14, 2014.

283 **4 billion screens lit today:** Calculation based on the following: Charles Arthur, "Future Tablet Market Will Outstrip PCs—and Reach 900m People, Forrester Says," *Guardian*, August 7, 2013; Michael O'Grady, "Forrester Research World Tablet Adoption Forecast, 2013 to 2018 (Global), Q4 2014 Update," Forrester,

December 19, 2014; and "Smartphones to Drive Double-Digit Growth of Smart Connected Devices in 2014 and Beyond, According to IDC," IDC, June 17, 2014.

283 **50 billion devices on the internet by 2020:** "Connections Counter," Cisco, 2013.

283 **another 13 billion appliances:** "Gartner Says 4.9 Billion Connected 'Things' Will Be in Use in 2015," Gartner, November 11, 2014.

283 **built into connected cars:** Ibid.

285 **6 billion times per year:** "$4.11: A NARUC Telecommunications Staff Subcommittee Report on Directory Assistance," National Association of Regulatory Utility Commissioners, 2003, 68.

285 **two lookups per week in the 1990s:** Peter Krasilovsky, "Usage Study: 22% Quit Yellow Pages for Net," Local Onliner, October 11, 2005.

285 **1 billion library visits per year:** Adrienne Chute, Elaine Kroe, Patricia Garner, et al., "Public Libraries in the United States: Fiscal Year 1999," NCES 200230, National Center for Education Statistics, U.S. Department of Education, 2002.

285 **$82 billion business:** Don Reisinger, "For Google and Search Ad Revenue, It's a Glass Half Full," CNET, March 31, 2015.

285 **four questions per day online:** Danny Sullivan, "Internet Top Information Resource, Study Finds," Search Engine Watch, February 5, 2001.

286 **ordinary people might pay for search:** Yan Chen, Grace YoungJoo, and Jeon Yong-Mi Kim, "A Day Without a Search Engine: An Experimental Study of Online and Offline Search," University of Michigan, 2010.

286 **average value of answering a question:** Hal Varian, "The Economic Impact of Google," video, Web 2.0 Expo, San Francisco, 2011.

INDEX

accelerometers, 221
accessing and accessibility, 109–33
  and clouds, 125–31
  and communications, 125
  and decentralization, 118–21, 125, 129–31
  and dematerialization, 110–14, 125
  and emergence of the "holos," 293–94
  as generative quality, 70–71
  ownership vs., 70–71
  and platform synergy, 122–25
  and real-time on demand, 114–17
  and renting, 117–18
  and right of modification, 124–25
accountability, 260–64
Adobe, 113, 206
advertising, 177–89
aggregated information, 140, 147
Airbnb, 109, 113, 124, 172
algorithms and targeted advertising, 179–82
Alibaba, 109
Amazon
  and accessibility vs. ownership, 109
  and artificial intelligence, 33
  cloud of, 128, 129
  and on-demand model of access, 115
  as ecosystem, 124
  and filtering systems, 171–72

and recommendation engines, 169
and robot technology, 50
and tracking technology, 254
and user reviews, 21, 72–73
anime, 198
annotation systems, 202
anonymity, 263–64
anthropomorphization of technology, 259
Apache software, 69, 141, 143
API (application programming interface), 23
Apple, 1–2, 123, 124, 246
Apple Pay, 65
Apple Watch, 224
Arthur, Brian, 193, 209
artificial intelligence (AI), 29–60
  ability to think differently, 42–43, 48,
    51–52
  as accelerant of change, 30
  as alien intelligence, 48
  in chess, 41–42
  and cloud-based services, 127
  and collaboration, 273
  and commodity consumer
    attention, 179
  and complex questions, 47
  concerns regarding, 44
  and consciousness, 42

artificial intelligence (AI) (*cont.*)
corporate investment in, 32
costs of, 29, 52–53
data informing, 39
and defining humanity, 48–49
and digital storage capacity, 265,
266–67
and emergence of the "holos," 291
as enhancement of human intelligence,
41–42
and filtering systems, 175
of Google, 36–37
impact of, 29
learning ability of, 32–33, 40
and lifelogging, 251
networked, 30
and network effect, 40
potential applications for, 34–36
questions arising from, 284
specialized applications of, 42
in tagging book content, 98
technological breakthroughs
influencing, 38–40
ubiquity of, 30, 33
and video games, 230
and visual intelligence, 203
See also robots
arts and artists
artist/audience inversion, 81
and augmented reality, 232
and authenticity, 70
and creative remixing, 209
and crowdfunding, 156–61
and low-cost reproduction, 87
and patronage, 72
public art, 232
attention, 168–69, 176, 177–89
audience, 88, 148–49, 155, 156–57
audio recording, 249. *See also* music and
musicians
augmented reality (AR), 216–17, 224,
226–27, 231–32
authenticity, 70
authority, 86, 88, 101
authors, 86, 87, 88
automation, 49–50, 55, 56, 57–58
automobiles. *See* transportation
avatars
and filtering systems, 175

and virtual reality technology, 212, 214,
217, 218–19, 232–33, 234
and virtual shopping, 173

Bailenson, Jeremy, 234–35
Barlow, John Perry, 138
Battlestar Galactica (series), 206, 282
Baxter, 51–53, 58
Baylor College, 225
Beats, 169
becoming, 9–27
and emergence of user-generated
content, 19, 21–22
and nascency of internet, 26–27
our blindness to, 13–22
and protopian narrative, 13–14
and technology-spawned
discontentment, 11–12
and upgrading, 10–11
Bell, Gordon, 247–48
Benkler, Yochai, 142
Bezos, Jeff, 111–12
Bing, 285
biofeedback, 225–26
biometrics and biodata, 235–36, 249, 263
Bitcoin, 120–21
BitTorrent, 66
blockbuster films, 196–97, 204, 208
blockchain technology, 120–21
blogs, 63, 89, 149
blood factor tracking, 238, 244
books
cognitive aspects of, 104
as conceptual state of imagination, 91
and consumer attention, 103, 178
culture of, 86–87, 88, 90
definition of, 90–91
durability of, 100–101
filtering superabundance of choices, 168
fixity of, 78–79
and immediacy of hardcovers, 68
impact of mass-produced, 85–86
included in the universal library, 102
and literacy techniques and
innovations, 200
and reader reviews, 72–73
and rewindability, 204
scanning of, 207

and tracking technology, 254
See also ebooks and readers
brain-machine interfaces (BMIs), 225
brands and branding, 167, 184
Brin, David, 260
Brooks, Rodney, 51, 53–54
Bush, Vannevar, 18, 19

caller identification, 253, 263
Call of Duty, 227
cameras, 221, 252
Carlsen, Magnus, 41–42
Carr, Nick, 78
car tracking. *See* transportation
Casablanca (1942), 202
celebrities, 187–88
censorship, 175–76
centaurs, 41
change, 5–7, 13–22, 30
Chardin, Teilhard de, 292
chess and artificial intelligence (AI),
   41–42
children and technology, 223
China, 4, 32, 56
cinematography, 198–200
Cisco, 283
civic duties, 271–72
clan-based societies, 262
"click dreaming," 280
clothing, 35, 224–25
clouds, 65, 125–31
code, 88, 90
collaboration, 141–42
   and digital socialism, 146
   and emergence of the "holos," 291
   and filtering systems, 171, 172
   and global connectivity, 276
   and increasing degree of
      coordination, 138
   and open source projects, 143
   and social impact of connectivity, 273
   and Wikipedia, 269–72
collectivism, 142–44, 151–52, 270–71
commercials, 197. *See also* advertising
commodity attention, 177–79
commodity prices, 189
communications
   and decentralization, 118–19, 129–31
   and dematerialization, 110–11

and free markets, 146
inevitable aspects of, 3
oral communication, 204
and platforms, 125
complexity and digital storage capacity,
   265–66
computers, 128, 231
connectivity, 276, 292, 294–95
consumer data, 256
content creation
   advertisements, 184–85
   custom music, 77
   early questions about, 17
   and editors, 148–51, 152, 153
   and emergence of user-generated
      content, 19, 21–22, 184–85,
      269–74, 276
   and Google search engines, 146–47
   and hierarchical/nonhierarchical
      infrastructures, 148–54
   impulse for, 22–23
   and screen culture, 88
   and sharing economy, 139
   value of, 149
convergence, 291, 296
cookies, 180, 254
cooperation, 139–40, 146, 151
copper prices, 189
copying digital data
   and copy protection, 73
   and creative remixing, 206–9
   and file sharing sites, 136
   free/ubiquitous flow of, 61–62, 66–68,
      80, 256
   generatives that add value to, 68–73
   and reproductive imperative, 87
   and uncopiable values, 67–68
copyright, 207–8
corporate monopolies, 294
coveillance, 259–64
Cox, Michael, 286–87
Craigslist, 145
Creative Commons licensing, 136, 139
crowdfunding, 156–61
crowdsourcing, 185
Cunningham, Ward, 135–36
curators, 150, 167, 183
customer support, 21
cyberconflict, 252, 275

dark energy and matter, 284
"dark" information, 258
Darwin, Charles, 243
data analysis and lifelogging, 250–51
"database cinema," 200
data informing artificial intelligence, 39, 40
decentralization, 118–21
    and answer-generating technologies, 289
    and bottom-up participation, 154
    and collaboration, 142, 143
    of communication systems, 129–31
    and digital socialism, 137
    and emergence of the "holos," 295
    and online advertising, 182–85
    and platforms, 125
    and startups, 116–17
    and top-down vs. bottom-up
        management, 153
Deep Blue, 41
deep-learning algorithms, 40
DeepMind, 32, 37, 40
deep reinforcement machine learning, 32–33
dematerialization, 110–14, 125, 131
diagnoses and diagnostic technology,
    31–32, 239, 243–44
diaries and lifelogging, 248–49
Dick, Philip K., 255
diet tracking, 238
Digg, 136, 149
digitization of data, 258
directional sense, 243
discoverability, 72–73, 101
DNA sequencing, 69
documentaries, updating of, 82
domain names, 25–26
Doritos, 185
Downton Abbey (series), 282
drones, 227, 252
Dropbox, 32
drug research, 241
DVDs, 205
Dyson, Esther, 186

Eagleman, David, 225
e-banks, 254
eBay, 154, 158, 185, 263, 272, 274
ebooks and readers, 91–96
    and accessibility vs. ownership, 112
    advantages of, 93–95

bookshelves for, 100
fluidities of, 79
interconnectedness of, 95–96, 98,
    99–100, 101–2, 104
and just-in-time purchasing, 65
liquidity of, 93
tagging content in, 98
and tracking technology, 254
echo chambers, 170
economy, 21, 65, 67–68, 136–38, 193
ecosystems of interdependent products
    and services, 123–24
editors, 148–51, 152, 153
education, 90, 232
Einstein, Albert, 288
electrical outlets, 253
email, 186–87, 239–40
embedded technology, 221
embodiment, 71, 224
emergent phenomena, 276–77, 295–97
emotion recognition, 220
employment and displaced workers,
    49–50, 57–58
Eno, Brian, 221
entertainment costs, 190
epic failures, 278
e-retailers, 253
etiquette, social, 3–4
evolution, 247
e-wallets, 254
experience, value of, 190
expertise, 279
exports, U.S., 62
extraordinary events, 277–79
eye tracking, 219–20

Facebook
    and aggregated information, 147
    and artificial intelligence, 32, 39, 40
    and "click-dreaming," 280
    cloud of, 128, 129
    and collaboration, 273
    and consumer attention system, 179, 184
    and creative remixing, 199, 203
    face recognition of, 39, 254
    and filtering systems, 170, 171
    flows of posts through, 63
    and future searchability, 24
    and interactivity, 235

and intermediation of content, 150
and lifestreaming, 246
and likes, 140
nonhierarchical infrastructure of, 152
number of users, 143, 144
as platform ecosystem, 123
and sharing economy, 139, 144, 145
and tracking technology, 239–40
and user-generated content, 21–22,
   109, 138
facial recognition, 39, 40, 43, 220, 254
fan fiction, 194, 210
fear of technology, 191
Felton, Nicholas, 239–40
Fifield, William, 288
films and film industry, 196–99, 201–2
filtering, 165–91
   and advertising, 179–89
   differing approaches to, 168–75
   filter bubble, 170
   and storage capacity, 165–67
   and superabundance of choices,
      167–68
   and value of attention, 175–79
findability of information, 203–7
firewalls, 294
first-in-line access, 68
first-person view (FPV), 227
fitness tracking, 238, 246, 255
fixity, 78–81
Flickr, 139, 199
Flows and flowing, 61–83
   and engagement of users, 81–82
   and free/ubiquitous copies, 61–62,
      66–68
   and generative values, 68–73
   move from fixity to, 78–81
   in real time, 64–65
   and screen culture, 88
   and sharing, 8
   stages of, 80–81
   streaming, 66, 74–75, 82
   and users' creations, 73–74, 75–78
fluidity, 66, 79, 282
food as service (FaS), 113–14
footnotes, 201
411 information service, 285
Foursquare, 139, 246
fraud, 184

freelancers (prosumers), 113, 115, 116–17,
   148, 149
Freeman, Eric, 244–45
fungibility of digital data, 195
future, blindness to, 14–22

Galaxy phones, 219
gatekeepers, 167
Gates, Bill, 135, 136
gaze tracking, 219–20
Gelernter, David, 244–46
General Electric, 160
generatives, 68–73
genetics, 69, 238, 284
Gibson, William, 214
gifs, 195
global connectivity, 275, 276, 292
gluten, 241
GM, 185
goods, fixed, 62, 65
Google
   AdSense ads, 179–81
   and artificial intelligence, 32, 36–37, 40
   book scanning projects, 208
   cloud of, 128, 129
   and consumer attention system, 179, 184
   and coveillance, 262
   and facial recognition technology, 254
   and filtering systems, 172, 188
   and future searchability, 24
   Google Drive, 126
   Google Glass, 217, 224, 247, 250
   Google Now, 287
   Google Photo, 43
   and intellectual property law, 208–9
   and lifelogging, 250–51, 254
   and lifestreaming, 247–48
   and photo captioning, 51
   quantity of searches, 285–86
   and smart technology, 223–25
   translator apps of, 51
   and users' usage patterns, 21, 146–47
   and virtual reality technology,
      215, 216–17
   and visual intelligence, 203
government, 167, 175–76, 252, 255, 261–64
GPS technology, 226, 274
graphics processing units (GPU), 38–39, 40
Greene, Alan, 31–32, 238

grocery shopping, 62, 253
Guinness Book of World Records, 278

hackers, 252
Hall, Storrs, 264–65
Halo, 227
Hammerbacher, Jeff, 280
hand motion tracking, 222
haptic feedback, 233–34
harassment, online, 264
hard singularity, 296
Harry Potter series, 204, 209–10
Hartsell, Camille, 252
hashtags, 140
Hawking, Stephen, 44
health-related websites, 179–81
health tracking, 173, 238–40, 250
heat detection, 226
hierarchies, 148–54, 289
High Fidelity, 219
Hinton, Geoff, 40
historical documents, 101
hive mind, 153, 154, 272, 281
Hockney, David, 155
Hollywood films, 196–99
holodeck simulations, 211–12
HoloLens, 216
the "holos," 292–97
home surveillance, 253
HotWired, 18, 149, 150
humanity, defining, 48–49
hyperlinking
    artifacts highlighted by, 279
    of books, 95, 99
    of cloud data, 125–26
    and creative remixing, 201–2
    early theories on, 18–19, 21
    and Google search engines,
        146–47

IBM, 30–31, 40, 41, 128, 287
identity passwords, 220, 235
IMAX technology, 211, 217
implantable technology, 225
indexing data, 258
individualism, 271
industrialization, 49–50, 57
industrial revolution, 189
industrial robots, 52–53

information production, 257–64. *See also*
    content creation
innovation, competitions for, 160
Instagram, 21, 139, 143, 199
instruction services, 69
Intel, 32
intellectual property, 207–8
intelligence, 41–47. *See also* artificial
    intelligence (AI)
interaction, 211–36
    costs of, 236
    and depth of attention, 282
    evolution of virtual worlds, 218–19
    and human senses, 219–27
    and mass customization, 173
    and "presence," 216–17
    and screening, 8
    social effects of virtual reality, 234–35
    state of current technology, 211–15
    and video games, 227–34
interactive devices, 253. *See also* virtual
    reality
Internal Revenue Service (IRS), 254
International Monetary Fund (IMF), 189
internet
    blindness to evolution of, 15–22
    commercialization of, 17–18
    and consumer attention, 177–78
    and copying digital data, 62
    creation of content on, 19, 21–22
    demographics for users of, 23
    and digital socialism, 137
    early expectations for, 15–16
    early worries about, 23
    and emergence of the holos, 293–94
    hyperlinked architecture of, 18–19, 21,
        146–47
    inevitable aspects of, 3
    nascent stage of, 26–27
    and participation of users, 22–23
    as public commons, 122
    self-policing culture of, 21
    and sharing economy, 144
    view of humanity from, 20
    See also content creation; web
internet of things, 175, 251, 283, 287
interpretation, 69
invention and inventiveness, 275
iPads, 155, 223, 224

iPhones, 123
Iron Man (2008), 222
It's a Wonderful Life (1946), 201
iTunes, 123, 266

Jefferson, Thomas, 207–8
just-in-time purchasing, 64–65

Kahle, Brewster, 96–97
Kaliouby, Rana el, 220
Kapor, Mitch, 151
Kasparov, Garry, 41
Keeley, Larry, 148
Kickstarter, 156–57
Kindle, 254
Kiva, 159

Lanier, Jaron, 213–14, 215, 219, 234
law and legal systems
    AI applications in field of, 55
    books of, 88, 90
    and clouds, 130
    code compared to, 88
    and creative remixing, 208
    and surveillance systems, 207
Leary, Tim, 214
Li, Fei-Fei, 203
libertarianism, 271
libraries
    Library of Everything, 165, 166–67, 190
    and printed books, 100–101
    public libraries, 86
    and tracking technology, 254
    universal library, 96–99, 101, 102
lifelogging, 105–6, 207, 246–50
lifestreaming, 244–51
light field projection, 216–17
LinkedIn, 32, 169
Linux operating system, 141, 143, 151, 273
liquidity, 66–67, 73–77, 88, 93. *See also* Flows and flowing
literacy, 86, 89, 90, 200–202
Local Motors, 160–61
location tracking, 226, 238, 243
Lost (series), 206, 282
Lucas, George, 198
luxury entertainment, 190
Lyft, 62, 252

machine intelligence, 266, 291. *See also* artificial intelligence
"machine readable" information, 267
Magic Leap, 216
malaria, 241
Malthus, Thomas, 243
Mann, Steve, 247
Manovich, Lev, 200
manufacturing, robots in, 52–53, 55
maps, 272
mathematics, 47, 239, 242–43
The Matrix (1999), 211
maximum likelihood estimation (MLE), 265
McDonalds, 25–26
McLuhan, Marshall, 63, 127
media fluency, 201
media genres, 194–95
medical technology and field
    AI applications in, 31, 55
    and crowdfunding, 157
    and diagnoses, 31
    future flows of, 80
    interpretation services in field of, 69
    and lifelogging, 250
    new jobs related to automation in, 58
    paperwork in, 51
    personalization of, 69
    and personalized pharmaceuticals, 173
    and pooling patient data, 145
    and tracking technology, 173, 237, 238–40, 241–42, 243–44, 250
Meerkat, 76
memory, 245–46, 249
messaging, 239–40
metadata, 258–59, 267
microphones, 221
Microsoft, 122–23, 124, 216, 247
minds, variety of, 44–46
Minecraft, 218
miniaturization, 237
Minority Report (2002), 221–22, 255
MIT Media Lab, 219, 220, 222
money, 4, 65, 119–21
monopolies, 209
mood tracking, 238
Moore's Law, 257
movies, 77–78, 81–82, 168, 204–7
Mozilla, 151
MP3 compression, 165–66

music and musicians
    AI applications in, 35
    creation of, 73–76, 77
    and crowdfunding, 157
    and free/ubiquitous copies, 66–67
    and intellectual property issues, 208–9
    and interactivity, 221
    liquidity of, 66–67, 73–78
    and live performances, 71
    low-cost reproduction of, 87
    of nonprofessionals, 75–76
    and patronage, 72
    sales of, 75
    soundtracks for content, 76
    total volume of recorded music, 165–66
Musk, Elon, 44
mutual surveillance ("coveillance"),
    259–64
MyLifeBits, 247

Nabokov, Vladimir, 204
Napster, 66
The Narrative, 248–49, 251
National Geographic, 278
National Science Foundation, 17–18
National Security Agency (NSA), 261
Nature, 32
Negroponte, Nicholas, 16, 219
Nelson, Ted, 18–19, 21, 247
Nest smart thermostat, 253, 283
Netflix
    and accessibility vs. ownership, 109
    and crowdsourcing programming, 160
    and on-demand access, 64
    and recommendation engines,
      39, 154, 169
    and reviews, 73, 154
    and sharing economy, 138
    and tracking technology, 254
Netscape browser, 15
network effect, 40
neural networks, 38–40
newbies, 10–11, 15
new media forms, 194–95
newspapers, 177
Ng, Andrew, 38, 39
niche interests, 155–56
nicknames, 263
nondestructive editing, 206

nonprofits, 157
noosphere, 292
Northwestern University, 225
numeracy, 242–43
Nupedia, 270

OBD chips, 251, 252
obscure or niche interests, 155–56
office settings, 222. *See also* work
    environments
on-demand expectations, 64–65, 114–17
OpenOffice, 151
open source industry, 135, 141–42, 143, 271
oral communication, 204
Oscar Awards, 187–88
overfitting, 170
ownership, 112–13, 117–18, 121–22,
    124–25, 127, 138

Page, Larry, 36–37
Pandora, 169
parallel computation, 38–39, 40
passive archives, 249
passwords, 220, 235
patents, 283
PatientsLikeMe, 145
patronage, 71–72
PayPal, 65, 119–20, 124
pedometers, 238
peer-to-peer networks, 129–30, 184–85
Periscope, 76
"personal analytics" engine, 239
personalization, 68–69, 172–73, 175, 191,
    240–41, 261–62
pharmaceutical research, 241–42
pharmacies, 50
phase transitions, 294–97
phones
    automatic updates of, 62
    cameras in, 34
    and clouds, 126
    and decentralized communications,
      129–31
    and on-demand model of access, 114
    directories, 285
    and interactivity, 219
    lifespan of apps for, 11
    as reading devices, 91–92
    in rural China, 56

and self-tracking technology, 239–40
and tracking technology, 239–40,
    250, 253
and virtual reality technology, 215, 222
photography and images
    and artificial intelligence, 33–34
    and classic film production, 198–99
    and content recognition, 43, 203
    and Creative Commons licensing, 139
    democratization of, 77
    and digital storage capacity, 266
    and facial recognition, 39, 43
    flexible images, 204
    and Google Photo, 43
    and lifelogging, 248–49
    and new media genres, 195
    and photo captioning, 51
    and reproductive imperative, 87
    sharing of, 140
Picard, Rosalind, 220
Picasso, Pablo, 288
Pichai, Sundar, 37
Pine, Joseph, 172–73
Pinterest, 32, 136, 139, 140, 183
piracy, 124
placebo effect, 242
platform synergy, 122–25, 131
PlayStation Now, 109
porn sites, 202–3
postal mail, 253
postindustrial economy, 57
"presence," 216–17
printing, 85, 87. *See also* books
privacy, 124, 253, 255
processing speeds, 293
Progressive Insurance, 251
Project Jacquard, 225
Project Sansa, 218
property rights, 207–8
prosumers (freelancers), 113, 115, 116–17,
    148, 149
proxy data gathering, 255
public commons, 121–22
public key encryption, 260–61
publishing and publishers, 149
purchase histories, 169

Quantified Self Meetup groups, 238–40
Quantimetric Self Sensing, 247

quantum computing, 284
Quid, 32
Quinn, David, 17

Radiohead, 72
randomized double-blind trials, 242
reading, 89, 91–92, 94–95, 103–4. *See also*
    books; ebooks and readers
realism, 211–14, 216
real time, 66, 88, 104, 114–17, 131, 145
recommendation engines, 169
Red Dead Redemption, 227–30
Reddit, 136, 140, 143, 149, 264
Red Hat, 69
reference transactions, 285
relationship network analysis, 187
relativity theory, 288
remixing of ideas, 193–210
    and economic growth, 193–95
    and intellectual property issues, 207–10
    legal issues associated with, 207–10
    and reduced cost of creating content,
        196–97
    and rewindability, 204–7
    and visual media, 197–203
remixing video, 197–98
renting, 117–18
replication of media, 206–9
Rethink Robotics, 51
revert functions, 270
reviews by users/readers, 21, 72–73, 139, 266
rewindability, 204–7, 247–48, 270
RFID chips, 283
Rheingold, Howard, 148–49
ride-share taxis, 252
ring tones, 250
Ripley's Believe It or Not, 278
robots
    ability to think differently, 51–52
    Baxter, 51–52
    categories of jobs for, 54–59, 60
    and digital storage capacity, 265
    dolls, 36
    emergence of, 49
    industrial robots, 52–53
    and mass customization, 173
    new jobs related to, 57–58
    and personal success, 58–59
    personal workbots, 58–59

robots (*cont.*)
  stages of robot replacement, 59–60
  training, 52–53
  trust in, 54
Romer, Paul, 193, 209
Rosedale, Phil, 219
Rowling, J. K., 210
RSS feeds, 63
Rukeyser, Muriel, 103

Samsung, 215, 219
Santa Fe Institute, 193
Scanadu, 243
science, 98–99, 157
Scout self-tracking device, 243
screening and screens, 85–108
  and content creation, 88
  and cultural tension, 88–89
  culture of, 87–88
  eyeglasses as future of, 105
  growth of the internet of things, 283
  interactivity of, 8, 103–4
  and interconnectedness of content,
    95–96, 98, 99–100, 101–2, 104
  and rewindability of visual media,
    204–7
  self-tracking with, 105–6
  ubiquity of screens, 86, 103
  and universal library, 96–99, 101, 102
  and virtual reality technology, 217
  See also ebooks and readers
"Screen of All Knowledge," 279
scroll-back functions, 206
search engines/browsers, 254, 286–88
Second Life, 218–19
security, 220, 235
self-filtering, 167
self-measurement, 237. *See also*
  lifestreaming
self-tracking, 105–6, 237–38, 266. *See also*
  lifestreaming
sensor technology, 226, 230, 267
Sensory Substitution Vest, 225
services, 6–7, 54, 62, 111–14, 125
sex, virtual, 218–19
sharing, 135–64
  and aggregated information, 140, 147
  and collaborative commenting sites, 136
  and consumers as content creators, 19

and crowdfunding, 156–61
and digital socialism, 136–38
and flowing, 8
and hierarchical/nonhierarchical
  infrastructures, 148–54
and increasing degree of coordination,
  138–42
motivation behind, 144
and obscure or niche interests, 155
and open source industry, 135, 141–42,
  143, 271
power of, 146
and social impact of connectivity, 271–75
societal problems addressed through, 146
technology facilitating, 145–46
ubiquity of, 139
Shirky, Clay, 138, 148
Sidecar, 62
Simon, Herbert, 176
singularity, 295–97
Siri, 287
sleep tracking, 238, 240
Smarr, Larry, 238
smart technology, 225, 253
smell, sense of, 226
Snapchat, 63, 199
Snowden, Edward, 261
socialism, digital, 136–38, 139, 146
social media
  and digital storage capacity, 265
  influence of, on public conversation, 140
  and intermediation of content, 150
  and privacy vs. transparency, 262
  and tracking technology, 254
  value of content created on, 149
  video and sound in, 76–77
social networks, 170, 186–87
software
  open source software, 135, 141–42, 143
  as service (SaS), 113
The Sopranos (series), 282
SoundCloud, 75
Spielberg, Steven, 221
Spotify, 74–75, 109, 138, 169, 254
Square, 65, 124
Squid (smart shirt), 225
Stanford Artificial Intelligence
  Laboratory, 203
Star Trek, 211

startups, 27, 33, 116–17, 184
Star Wars films, 198–99
statistics, 242–43
stickiness, 69
Stoll, Cliff, 15–16
storage capacity, 166, 264–67
streaming devices, 254. *See also*
    lifestreaming
StumbleUpon, 136, 145
subscription services, 112–13, 114
superabundance, 168–69, 176, 189
Super Bowl, 183, 185
superlatives, 277–79
surfing the web, 188–89, 280–82
surveillance, 230, 252–53, 255, 259–64
survival skills, 243
sustainable growth, 193
Swift, Taylor, 75
synthetic senses, 242–44
synthetic worlds, 219, 230

tags, 63, 98, 139, 140
technium, 273–74
TED videos, 278
television, 88–89, 177, 223, 282
texting, 89
textual literacy, 198, 201–2
thermostats, 223, 253, 283
3-D modeling, 199–200, 211–12, 232
3-D printers, 173
3-D Warehouse, 141
time, 64. *See also* real time
Time Machine backup system, 246
TiVos, 205
Toffler, Alvin, 113, 148
top-down vs. bottom-up management,
    148–54
total lifetime spending, 187
total recall, 248
tracking, 236–67
    bias toward tracking, 252–57
    and cognifying, 8
    and consumer attention system, 184
    and coveillance, 259–64
    current practices and trends, 237–42
    and digital storage capacity, 264–67
    and growth of connected devices, 283
    and lifestreaming, 244–51
    and mass customization, 173

and platform-selected advertising, 183
    and rate of information production,
        257–64
    routine tracking, 255
    self-tracking, 105–6, 237–38, 266 (*see
        also* lifestreaming)
    and synthetic senses, 242–44
    and virtual reality technology, 219
traffic monitoring, 252
trailers, 202
transhuman minds, 44
translators, 51, 104–5
Transparent Society, 260–61
transportation, 43, 57–58, 62, 111–12,
    115–16, 252. *See also* Uber
travel data, 239–40
trust, 67, 264
Tumblr, 136, 139
TV Guide, 72
Twitter and tweeting
    and aggregated information, 147
    and anonymity, 263–64
    and artificial intelligence, 32
    and creative remixing, 194
    and etiquette, 3
    and filtering systems, 169, 170
    and hashtags, 140
    influence of, on public conversation, 140
    and self-tracking technology, 239–40
    sharing information on, 145
    as streaming technology, 63
    and user-generated content, 21, 138

Uber
    on-demand services of, 62, 114
    and filtering systems, 172
    model of, 115
    peer-to-peer networking, 183–84
    and social impact of connectivity, 273, 274
    success of, 154
    and tracking technology, 252
Uncharted 2, 227
Underkoffler, John, 222
upgrading technology, 10, 62–63
US Supreme Court, 270

Varian, Hal, 286
video and video technology
    ease of creating videos, 166

video and video technology (*cont.*)
  and filtering systems, 196–99
  and lifelogging, 249
  and new media fluency, 201–3
  and rewindability, 204–7
  streaming, 205
  and user-created content, 82
video games and industry
  and artificial intelligence, 32–33, 230
  and creative remixing, 195
  and depth of content, 282
  and graphics processing units
    (GPU), 38
  interactivity of, 103, 227–34
  narrative in virtual reality games, 229
  and rewindability, 206
  and virtual reality technology, 215–16
vigilantes, 263
Vine, 76, 194
virtual reality
  and computing practices, 222–23
  and digital storage capacity, 265
  and immersion, 226–27
  and interactivity, 211–15, 218–27
  revolutionary nature of, 231
  social effects of, 234–35
  varied uses for, 229
  VR headsets, 219

Wachter, Udo, 243
Warhol, Andy, 209
Watson, 30–31, 40, 287
wearable devices
  growth of industry, 283
  and interactivity, 224–25
  and lifelogging, 248, 251
  and lifestreaming, 244
  and rewindability, 207
  and synthetic senses, 243–44
web
  anticipation of users' needs, 25
  blindness to evolution of, 15–22
  and context of time, 24–25
  and fear of commercialization, 17–18
  and future searchability, 24
  genesis of, 19
  and Google search engines, 146–47
  hyperlinked architecture of, 18–19, 21,
    146–47

page content on, 89
surfing, 188–89, 280–82
and tracking technology, 254
ubiquity of, 25
and website design, 220
See also internet
WeChat, 63, 76, 124, 246
Weiswasser, Stephen, 16
Wells, H. G., 292
WhatsApp, 63, 76, 199
whistle-blowers, 261, 263–64
Wikipedia
  and artificial intelligence, 39
  collaborative efforts of, 135–36
  and collective content creation,
    143, 269–74, 276
  correcting errors in, 104
  and Creative Commons licensing, 136
  dynamic content in, 102
  and editors, 151, 152, 153–54
  hybrid model of, 144
  interlinking in, 95–96, 98
  and nondestructive editing, 206
  and screen culture, 93
"wild" information, 258
The Wire (series), 206, 282
Wired magazine, 16–17, 18, 25–26,
    148–49, 238
The Wizard of Oz (1939), 203
Wolf, Gary, 238, 250
Wolfram, Stephen, 239
work environments, 217, 222, 232–33, 280

X-Ray vision, 226, 231

Yahoo! 32, 170, 285
Yelp, 139, 266
YouTube
  and convergence of internet media, 282
  and creative remixing, 196–97, 201
  and improbable events, 277
  number of users, 143
  and sharing economy, 139
  and tracking technology, 254
  and ubiquity of unlikely events, 277
  and user-created content, 21–22, 273

"zillion" term, 264–65, 276
Zip, 62